P9-CDK-600

75
FSG

ALSO BY GEOFF MANAUGH

A Burglar's Guide to the City

UNTIL PROVEN SAFE

UNTIL PROVEN SAFE

THE HISTORY AND FUTURE OF QUARANTINE

GEOFF MANAUGH AND **NICOLA TWILLEY**

MCD | FARRAR, STRAUS AND GIROUX | NEW YORK

MCD
Farrar, Straus and Giroux
120 Broadway, New York 10271

Copyright © 2021 by Geoff Manaugh and Nicola Twilley
Maps copyright © 2021 by Jeffrey L. Ward
All rights reserved
Printed in the United States of America
First edition, 2021

Some of this material previously appeared, in substantially different form, in the following publications: Chapters 3, 4, 5, 7, and 8 include fragments of interviews and other materials previously published on *BLDGBLOG* (bldgblog.com). Chapters 4, 6, 8, and 9 include material previously published by *The New Yorker* (newyorker.com). Chapter 2 includes material previously published by *The New York Times* (nytimes.com).

The images that were made available by the Wellcome Collection are used under a Creative Commons license, Attribution 4.0 International (CC BY 4.0).

Library of Congress Cataloging-in-Publication Data
Names: Manaugh, Geoff, [date] author. | Twilley, Nicola, [date] author.
Title: Until proven safe : the history and future of quarantine / Geoff Manaugh and Nicola Twilley, MCD.
Description: First edition. | New York : MCD/Farrar, Straus and Giroux, 2021. | Includes bibliographical references and index.
Identifiers: LCCN 2021006853 | ISBN 9780374126582 (hardcover)
Subjects: LCSH: Quarantine—History. | Epidemics—History. | COVID-19 (Disease)—History.
Classification: LCC RA655 .M36 2021 | DDC 614.4/6—dc23
LC record available at https://lccn.loc.gov/2021006853

Our books may be purchased in bulk for promotional, educational, or business use. Please contact your local bookseller or the Macmillan Corporate and Premium Sales Department at 1-800-221-7945, extension 5442, or by email at MacmillanSpecialMarkets@macmillan.com.

www.mcdbooks.com • www.fsgbooks.com
Follow us on Twitter, Facebook, and Instagram at @mcdbooks

10 9 8 7 6 5 4 3 2 1

CONTENTS

SOUTHERN EUROPE

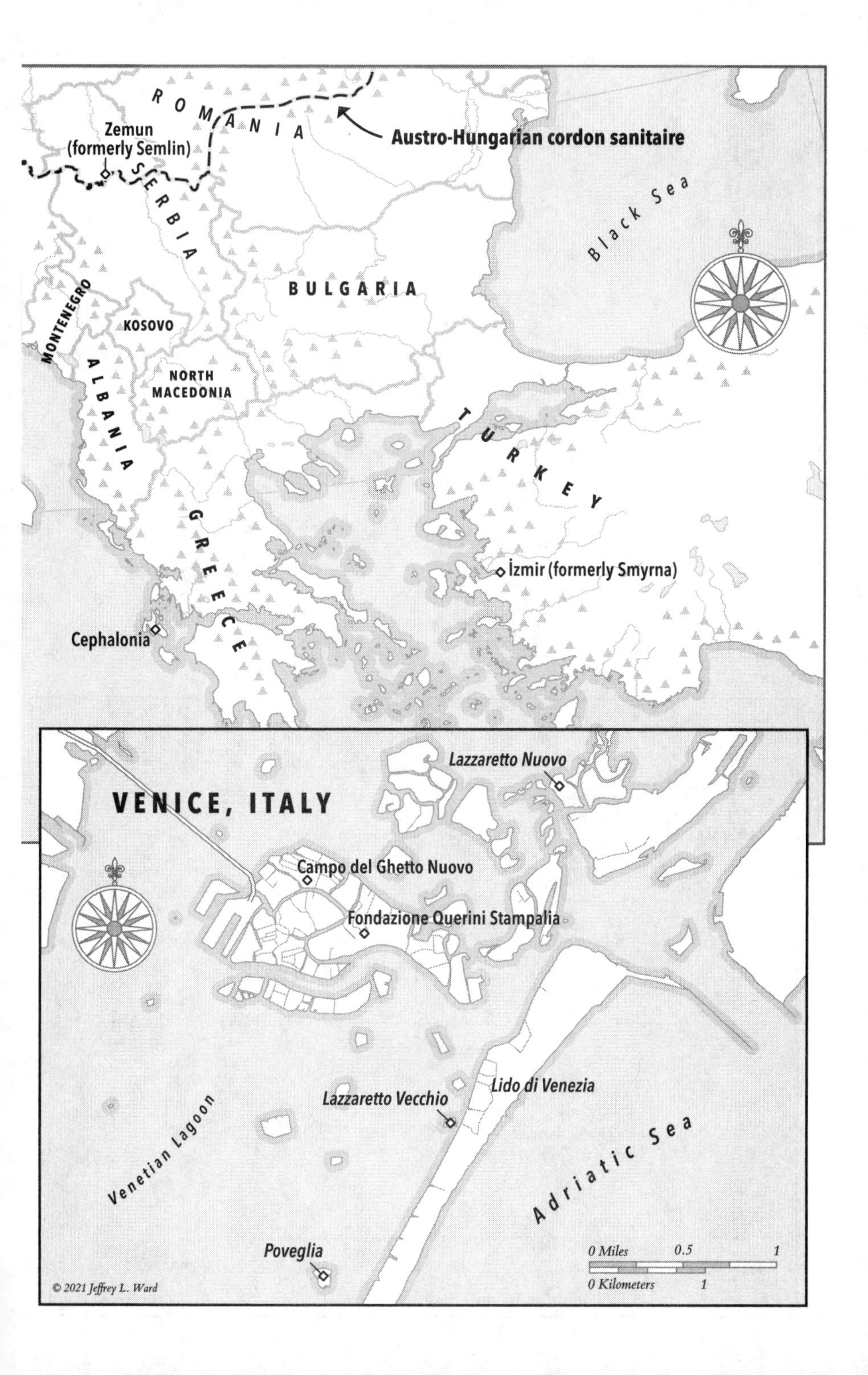
ROMANIA
Zemun
(formerly Semlin)
Austro-Hungarian cordon sanitaire
SERBIA
Black Sea
BULGARIA
MONTENEGRO
KOSOVO
ALBANIA
NORTH
MACEDONIA
TURKEY
GREECE
İzmir (formerly Smyrna)
Cephalonia
VENICE, ITALY
Lazzaretto Nuovo
Campo del Ghetto Nuovo
Fondazione Querini Stampalia
Lido di Venezia
Lazzaretto Vecchio
Venetian Lagoon
Adriatic Sea
Poveglia
0 Miles 0.5 1
0 Kilometers 1
© 2021 Jeffrey L. Ward

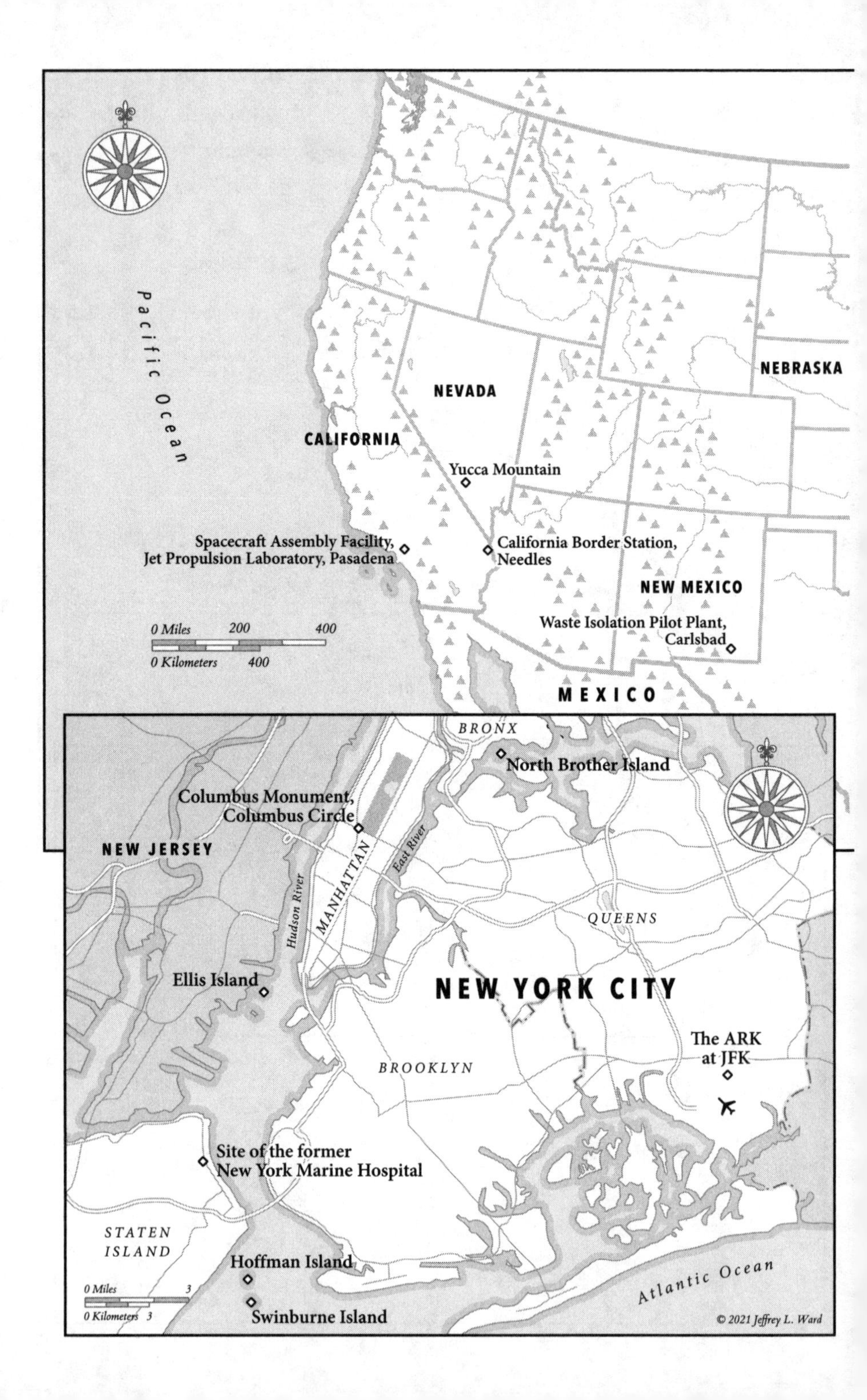

Pacific Ocean
NEVADA
CALIFORNIA
NEBRASKA
Yucca Mountain
Spacecraft Assembly Facility,
Jet Propulsion Laboratory, Pasadena
California Border Station,
Needles
NEW MEXICO
Waste Isolation Pilot Plant,
Carlsbad
0 Miles 200 400
0 Kilometers 400
MEXICO
BRONX
North Brother Island
Columbus Monument,
Columbus Circle
NEW JERSEY
MANHATTAN
East River
Hudson River
QUEENS
Ellis Island
NEW YORK CITY
The ARK
at JFK
BROOKLYN
Site of the former
New York Marine Hospital
STATEN
ISLAND
Hoffman Island
0 Miles 3
0 Kilometers 3
Swinburne Island
Atlantic Ocean
© 2021 Jeffrey L. Ward

THE UNITED STATES

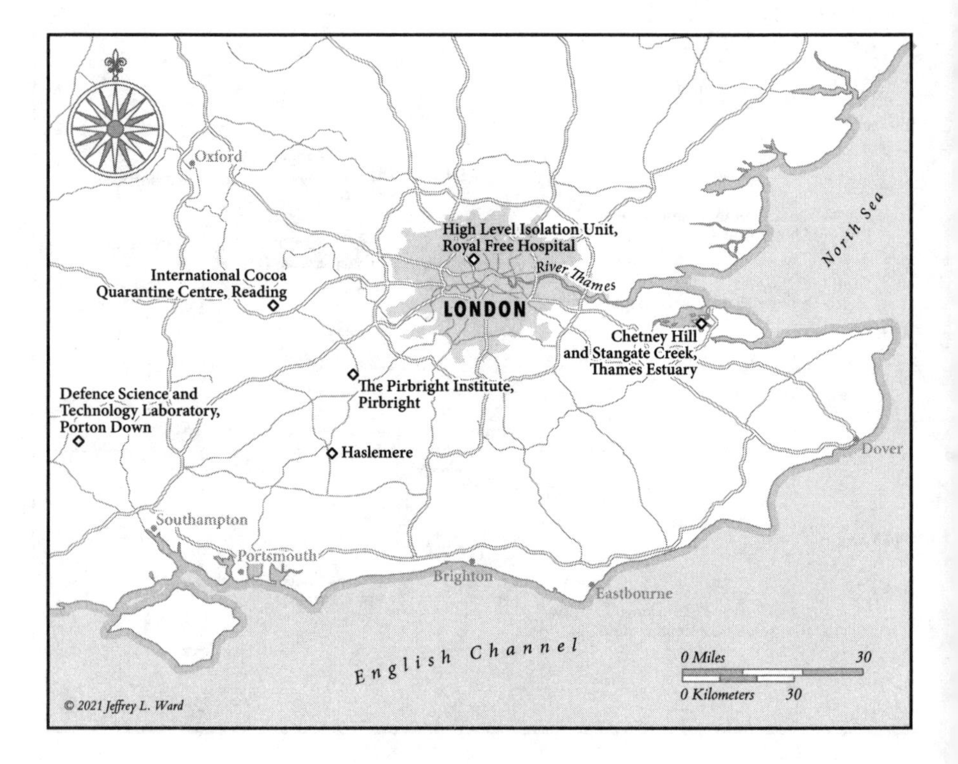

UNITED KINGDOM

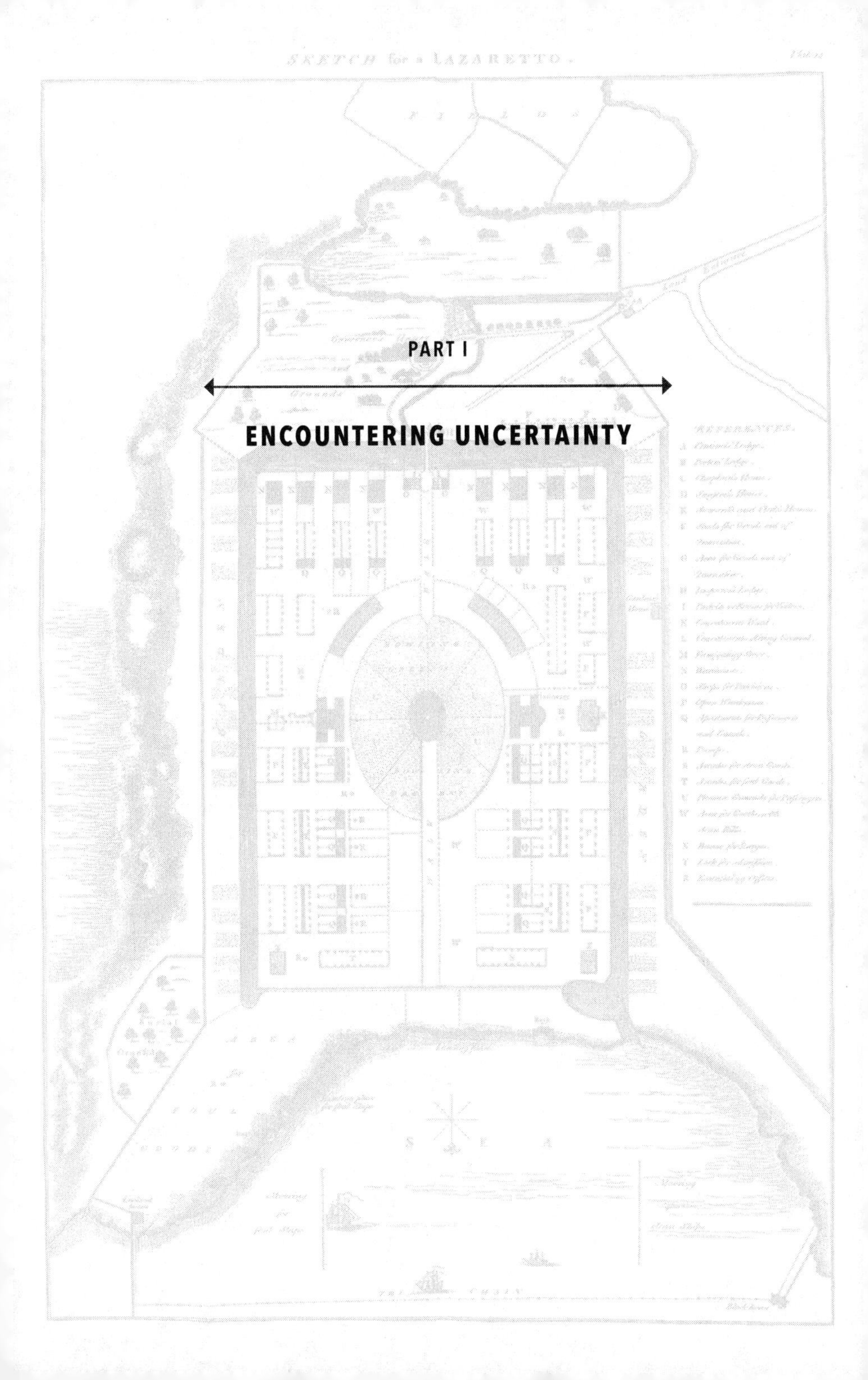

PART I

ENCOUNTERING UNCERTAINTY

1

The Coming Quarantine

On March 6, 2020, a King County, Washington, health department van pulled up in front of an Econo Lodge motel outside Seattle. An employee clad in white coveralls hopped out, grabbed tools from the back of his van, and proceeded to paint the motel's still-glowing sign pitch-black. The red and yellow colors of the chain's familiar logo quickly disappeared, replaced by a matte-black rectangle that loomed over the street like a pirate flag. Ominous, deathly, its former welcoming light now extinguished, the motel had become a quarantine facility.

The lo-fi nature of the motel's transformation was an unsettling indicator of just how improvised and ad hoc quarantine preparations seemed to be when a new infectious disease—known as COVID-19—first arrived in the United States. As this novel coronavirus spread exponentially and hospital beds filled up, public health officials realized that they had nowhere to put people who could not quarantine at home.

Instead, buildings such as this roadside motel—purchased by Washington State health officials for $4 million—were hastily retrofitted, becoming part of the nation's emergency medical infrastructure overnight. In this case, the motel's rooms were already equipped with independent HVAC units, doors that opened to the outside, and seamless, easy-clean floors. All it took to complete the transformation was a coat of black paint.

⟷

That same week, a friend of ours invited us to join an international preppers' list hosted on the encrypted communication app Telegram. The purpose of the list was allegedly to help its members prepare themselves and their families for what appeared to be an imminent national lockdown; there were tips for securing enough toilet paper, advice on baking bread, and tales of making a first-time handgun purchase.

In the many thousands of messages hosted by the group, posted by often-anonymous users from around the world, we saw a sign of things to come in terms of public perception of the coronavirus pandemic. Some members wondered aloud about imaginary connections between COVID-19 and 5G wireless technology; others began formulating a muddled conspiracy theory in which the billionaire Bill Gates planned to use a future vaccine to inject electronic nanoparticles into human subjects against their will. If there was any doubt that global health authorities had lost control of the information war before the pandemic had even really started, this Telegram group quickly dispelled it.

More striking to us than the group's embrace of misinformation and conspiracy theories was the profound, almost palpable fear of a coming quarantine. For members based in the United States—a country whose popular identity has been constructed around notions of freedom of movement and individual liberty—political fears of government overreach became fused with the morbid dread of a global plague. Every day, it seemed, the looming specter of quarantine crept closer, depicted as a dictatorship of doctors in which we would all be considered infectious until proven safe.

By this time, China was already several weeks into its own mass lockdown aimed at containing the novel coronavirus; tens of millions of people had been quarantined for potential exposure to COVID-19, entire cities forcibly isolated from the rest of the world. A popular view emerged in Western media during those early weeks that only an authoritarian government such as China's could even attempt such a thing.

Skepticism about U.S. quarantine capabilities was, in fact, warranted: as cases of the virus began to grow in number, the nation's permanent federal quarantine infrastructure consisted of just twenty inspection stations at international airports around the country and a brand-new, twenty-bed unit in Omaha, Nebraska. This, the nation's only federal quarantine facility, barely opened in time for COVID-19: after a lengthy construction process, it became operational on January 29, 2020. Nevertheless, nightly news reports and incendiary social media posts only confirmed for many members of this Telegram group that mass quarantine was imminent. Convinced that the government exercise of such extraordinary powers was potentially illegal and certainly un-American, they readied themselves to resist.

Even Chinese authorities seemed caught off guard by the scale of COVID-19. *The New York Times* described efforts to isolate or quarantine people in the city of Wuhan—where the disease is believed to have originated—as a "mass roundup" marked by "chaos and disorganization." Public health authorities were "haphazardly" gathering up "sick patients, in some cases separating them from their families." Family members and close contacts of confirmed cases were also dispatched into centralized quarantine and observation facilities—primarily existing buildings, including stadiums, convention centers, and schools that had undergone emergency conversion into strange new types of frontline medical infrastructure.

Among these were so-called fever buildings and *fangcang* hospitals. *Fangcang*—a name that in Chinese is a homophone for "Noah's ark"—were large temporary hospitals, often retrofitted sports and exhibition centers, to which people with mild or asymptomatic cases of COVID-19 could be sent, to isolate them from the community while supplying them with food, shelter, and social activities. Fever buildings were their dark cousins: entire housing complexes in which so many inhabitants had contracted the virus that authorities simply put a cordon around the entire structure. Large signs affixed outside warned healthy people to stay away. When converted facilities were not sufficient for the detention of tens of thousands of people suspected of exposure to COVID-19, however, China's famously efficient construction industry clicked into gear. In one case, a sprawling thousand-bed hospital made from modular architectural units was assembled in only ten days by workers continually checked for symptoms of coronavirus infection.

For the historically minded, these sights resonated with medical efforts of the past; the implementation of quarantine and isolation has always been a stimulus for creatively rethinking the built environment. For centuries, pandemic disease has inspired people to find new uses for old buildings or to invent new structures altogether. In sixteenth-century England, following ordinances proclaimed by King Henry VIII, the houses of people in quarantine had to be marked with long white poles attached to the exterior walls, like the quills of a porcupine, with clumps of straw or hay attached to the ends. These functioned as highly visible warning signs as well as inconvenient physical obstacles, encouraging pedestrians and carriages to avoid certain streets altogether.

In Venice, Italy, toward the end of the 1500s, the houses of the quarantined were also marked and labeled with prominent warning signs, including wooden crosses, then boarded up and locked from without to prevent potentially infected inhabitants from breaking free. Commentators at the time described feelings of horror as they gazed upon thousands of homes forcibly shuttered around the city, many of them with people still inside.

Families stuck behind the bright yellow plastic barriers that snaked through Wuhan, cutting neighborhoods off from one another, must have felt similarly trapped. For apprehensive members of our Telegram group, the scale and rapidity of this official Chinese response was not inspiring; it was frightening. Were mass roundups, quarantine camps, and forced hospitalizations on their way to the United States?

The group's mood grew darker as news emerged from beyond China's major population centers. Linking to stories both reported and conspiratorial, members noted that Chinese villagers had begun using heavy construction equipment and agricultural machines to block roads into and out of town, enforcing their own makeshift lockdowns. Reuters described these as "vigilante" quarantines, thrown together with caution tape and cinder blocks, more *Mad Max* than World Health Organization (WHO). Commuters, stranded in their cars, were forced to sleep behind the wheel, alone, with neither food nor access to their homes and families. As human mobility ground to a halt, even the circulation of money and mail was affected: the People's Bank of China instituted fourteen-day quarantines for banknotes from virus-stricken areas, while China Post announced it would quarantine envelopes and packages, delivering them only once transportation corridors could be reopened.

Then, just a couple of days after the Seattle Econo Lodge sign was painted black, the global dam seemed to break. Cases of COVID-19 began popping up everywhere, in South Korea, Iran, and Israel, in London and New York City.

Italy announced plans to lock down the entire region of Lombardy, in the country's wealthy north. In U.S. media, this was all but confirmation that, for whole swaths of the country, from the Pacific Northwest to Texas, the clock was ticking. "Italy has been plunged into chaos," *The Guardian* wrote, describing the country's attempt to quarantine sixteen million people overnight. TV footage of commuters sprinting through railway stations, attempting to catch the last trains out of town, desperate to reunite with their families, captured an anxiety bordering on terror. We would later learn that as many as thirty thousand students boarded trains, hoping to return to their families in Italy's south before the lockdown began. Upon arrival, these students were greeted by police-assisted teams of epidemiologists who ordered them directly into quarantine. "We know your name," the police said to each student, according to Dr. Luigi Bertinato, the lead scientific advisor to Italy's national COVID-19 response team. "We know where you live. We will check on you."

What such footage also showed is that, when faced with an imminent quarantine order, large portions of a targeted population are almost guaranteed to flee, sometimes carrying the disease with them, seemingly invalidating the purpose of the intended lockdown in the process. Several weeks earlier, as many as five million people had fled Wuhan in anticipation of lockdown measures; many of those five million went straight to crowded megacities, such as Beijing or Shanghai, or flew halfway around the world to destinations in Europe and North America, potentially seeding further outbreaks when they arrived. Quarantines are often compromised before they even begin: their threat alone can drive a disease underground, making its spread harder to track, let alone control.

Indeed, when the Trump administration abruptly announced an impending closure of the U.S. border to European arrivals in mid-March, it precipitated a reverse-rush of travelers back into the United States. Tens of thousands of people, afraid of being cut off from their homes and families for an indefinite period of time in the middle of a global pandemic, spent thousands of dollars apiece on emergency return tickets, sometimes canceling nonrefundable hotel reservations and other travel plans at enormous personal expense.

Writing in *The Washington Post* about her own experience, Cheryl Benard, a former health analyst for the RAND Corporation, described flying home to Washington, D.C., on a one-way flight from Vienna as "a case study in how to spread a pandemic." Benard found herself standing in a packed international-arrivals hall at Dulles airport for hours, with no space between travelers—let alone between long lines of people known to be infected with COVID-19. "When I asked a security guard about the other lines," Benard wrote, "he told me they were for people with a confirmed corona diagnosis. There was no separation for this group—no plastic sheets, not even a bit of distance. When your line snaked to the left, you were inches away from the infected."

To the horror of our Telegram group, the Italian government suspended parts of the country's constitution, enabling the expansion of quarantine orders to include the entire nation. Macabre details began to surface as the lockdown intensified. Among the eeriest of these was the story of Luca Franzese, a Neapolitan forced to quarantine at home with the corpse of his own sister, who had died of COVID-19. With local funeral homes refusing to take her body out of fear of contagion, Franzese found himself living alongside her remains for two horrific days. "Franzese posted an emotional appeal to his followers on Facebook," *The Washington Post* reported, "urging them to take the virus seriously as he stood in the same room where his sister lay dead in the background. 'We are ruined,' he said. 'Italy has abandoned us.'"

Equally chilling reports emerged of elderly and vulnerable populations locked at home without adequate food or medication, and women and children trapped indoors with only their abusers for company. "Many women under lockdown for #COVID19 face violence where they should be safest: in their own homes," António Guterres, the secretary-general of the United Nations, tweeted in early April. "Today I appeal for peace in homes around the world."

Meanwhile, with Chinese authorities reporting that their own outbreak seemed to be under control, Asians of all ethnicities living in Australia, Europe, and the United States found themselves shunned by their fellow citizens, banned from entering businesses, and accused of importing the coronavirus. These racist attacks ranged from verbal abuse—President Donald J. Trump insisted on referring to COVID-19 as the "Chinese virus," and, later, "kung flu"—to physical attacks. In Midland, Texas, a Burmese American father and his sons were stabbed while shopping at a Sam's Club; their attacker explained that he thought they were Chinese and that he was trying to stop them from spreading coronavirus in the community.

By late March, restaurants, cinemas, gyms, and schools across the United States began to close. Panic-shopping and toilet-paper hoarding kicked into high gear; fights broke out in supermarkets and big-box stores, requiring police intervention, with security guards forced to separate families battling over Clorox and Cottonelle. In Tennessee, a young man stockpiling hand sanitizer had his storage unit raided by police, his antibacterial hoard confiscated and donated to a needy public. Luxury quarantine schemes began to appear online, advertising accommodation in the California desert for people rich enough to escape an infected world. Even listings on mainstream travel websites, such as Travelocity and Airbnb, were hastily rewritten to entice people looking for a comfortable place to quarantine, now emphasizing a destination's cleanliness and distance from its neighbors. Unusual new phrases—"social distancing," "managed isolation"—became ubiquitous, describing efforts to avoid other people.

Here where we live in Los Angeles, temporary lines of colored tape appeared everywhere, marking safe places to sit or stand in public. The interiors of grocery stores were rearranged so that customers could maintain distance as they bulk-bought pasta. We found ourselves living under a stay-at-home order, marveling at the terrible irony of finishing a book about quarantine while in a state of medical lockdown.

Of course, none of these public health guidelines would have amounted to much if no one had been willing to enforce them. Sometimes, these efforts provided a surreal kind of comic relief: in late March, as the country entered its second week of lockdown, Italians circulated supercuts of their mayors berating them for leaving home. "Where are you going with these incontinent dogs?" yelled Massimiliano Presciutti, the mayor of the picturesque Umbrian hill town of Gualdo Tadino, in a video posted to Facebook. "People are dying, don't you get it?" Dog walking was one of the few activities for which Italians were allowed to go outside, a loophole that quickly resulted in an underground pet-rental economy and some seriously pooped pooches.

As economic and social disruption from the coronavirus grew, *The Washington Post* warned that "under all kinds of political systems, governments are turning to increasingly stringent measures—and deploying their armed forces to back them up." They were not referring to just *official* armed forces. In a Rio de Janeiro favela known as the City of God, gangs began imposing their own quarantine restrictions, ordering people to stay at home if they thought they or their families had been exposed. Brazil's president, Jair Bolsonaro—who would later test positive for the coronavirus—infamously refused to mount a national response to COVID-19, rejecting all scientific guidance around masks, economic shutdowns, and quarantines. Everyday Brazilians were left to fend for themselves.

In Japan, so-called virus vigilantes took quarantine into their own hands, dumping thumbtacks on the street to pop the tires of potentially infected cyclists attempting to bike through neighborhoods where they didn't belong, and scouring cities for cars with out-of-town registration details. The ominous, increasingly sci-fi outlines of a looming dystopia, in which individual liberties were subsumed by medico-political constraints, continued to emerge. China—then Italy, Spain, and France, to name but three—began buzzing law enforcement drones over the heads of residents ignoring lockdown orders. Drone-mounted speakers blared demands that people immediately turn around and go home. Food-delivery robots went from high-tech curiosity to pragmatic logistical infrastructure almost overnight, in the process offering a glimpse of the fully automated quarantines likely to come in the near future. In India, doctors began stamping mandatory stay-at-home orders directly onto people's forearms with indelible ink; for some commenters in our group, this was uncomfortably close to the Nazi use of tattoos to mark concentration camp detainees in World War II.

When an op-ed came out in *The Washington Post* soon thereafter demanding that Trump institute widespread spatial controls on the movements of American citizens—imploring, "Mr. President, lock us up!"—many members of our Telegram group felt not just betrayed but endangered. They seemed less fearful of the disease, we noted, than of the tactics that might be used to protect them from it, a response that would become only more common—and more life-threatening—as the long, terrible year of 2020 dragged on.

It should have come as no surprise by then that cases of the coronavirus would rapidly multiply throughout the United States. For all that they had seen unfold over the past few weeks in China—a lengthy head start during which they could have distributed personal protective equipment (PPE) and built a robust coronavirus-testing infrastructure—U.S. authorities were shockingly unprepared. In fact, an attempt to designate six major ports of entry as "sentinel cities," using enhanced surveillance to provide early warning of the disease's movements, failed almost immediately because tests produced and distributed by the Centers for Disease Control and Prevention (CDC) proved unreliable.

In the absence of any coherent direction from the Trump administration came public reliance on misinformation and widespread distrust of official guidance from the WHO or the CDC. Some of our Telegram list members claimed that, in Iran, families were being welded inside their own homes; the only proof of this brute-force quarantine was grainy, unsourced footage depicting someone using an acetylene torch at night (supposedly somewhere in Tehran). Other members insisted that, in China, married couples had begun committing suicide together, leaping to their deaths from apartment windows, driven mad by hunger and isolation. In these cases, too, unverified smartphone video clips lacking dates or geographic context were the only evidence. In one unfortunate instance, an image taken from a recent science fiction film was misidentified as a scene of authoritarian quarantine taking place on a highway somewhere outside Moscow.

The United States soon had its own virus vigilantes and makeshift quarantines. Toward the end of March, Rhode Island police set up inspection stations at the state's borders, *Bloomberg* reported, deputized "to hunt down New Yorkers" who might be carrying the disease, even authorizing "house-to-house searches" to locate any New Yorkers—infected or not—who might be trying to hide. In mid-March, Dare County, North Carolina—a picturesque archipelago off the Atlantic coast, home to Cape Hatteras National Seashore and the Outer Banks—announced it was cutting itself off from the outside world. As if its resemblance to the quarantine islands of medieval Venice were already not enough, Dare County officials borrowed another plague-era innovation: health documents. Residents traveling within Dare County had to display a permit on their vehicle dashboards, for some only adding to an atmosphere of medical imprisonment.

On another island—Vinalhaven, off the coast of Maine—summer-home owners fleeing COVID-19 arrived to find their presence was not welcome. In one case, hostile islanders armed with shotguns and chainsaws cut down a family's trees in order to block their driveway and seal them inside their home. Pop-up, impromptu, inspired more by Home

Depot than by the CDC, this must have seemed as good a way as any to prevent a mysterious disease from spreading—assuming, of course, this out-of-town family even had it. Without access to widespread testing, anyone could be infected and they—or their neighbors—would never know. Uncertainty was everywhere—thus the need for quarantine.

By the end of March 2020, a mere three weeks after King County health authorities painted over an Econo Lodge sign, fully 20 percent of the Earth's population was living in a state of isolation or quarantine—an estimated 1.7 billion people. Within a week, that tally would more than double, with Agence France-Presse estimating that "half of humanity" was now undergoing some form of medical detention. A global lockdown had arrived seemingly out of nowhere, shutting down the planet in weeks. Everyday life—normal interactions with family, colleagues, neighbors, and friends that people around the world took for granted—now seemed impossibly distant and dangerous, replaced by a bleak new world of stay-at-home orders, statewide lockdowns, and mandatory quarantine.

The very word—foreign, clinical, medieval—inspired fear.

⟷

Quarantine—from the Italian word *quarantena*, shorthand for *quaranta giorni*, meaning "forty days"—is one of humanity's oldest and most consistent responses to epidemic disease. In theory, quarantine has always been minimal—it works by separating people suspected of sickness from those known to be well—but making this seemingly simple distinction opens up entire worlds of philosophical uncertainty, ethical risk, and the potential abuse of political power.

While the successful implementation of quarantine can

be logistically challenging, the logic behind it remains straightforward: there might be something dangerous inside you—something contagious—on the verge of breaking free. The space and time you need to see whether it will emerge is quarantine. Quarantine can be an effective medical tool, but it is also an unusually poetic metaphor for any number of moral, ethical, and religious ills: it is a period of waiting to see if something hidden within you will be revealed.

For hundreds, if not thousands, of years, quarantine has been used to create a buffer between the known and the unknown; it delays our contact with something we do not understand and against which we have no natural immunity. It is an inherently spatial solution, with surveillance at its core. When architects and engineers design spaces of quarantine, whether those are purpose-built hospital wards, modified Airstream trailers, converted motels, or airtight greenhouses, they are attempting to remove risk from our interactions with others. For this exact reason, quarantine historically most often occurs at sites of encounter, at ports of entry, at places of meeting between one culture, even one species, and another. In fact, some of today's existing geopolitical borders are where they are because of quarantine: dividing lines that have persisted where nation-states and empires once grew nervous about protecting themselves from an approaching threat. If there is something dreadful, even potentially fatal, on its way, it is not surprising that we might want to find a way to delay its arrival. That delay is quarantine.

Today, quarantine is more relevant than ever, as we alter the environment in ways that have tipped the balance in favor of novel pathogens. Based on sheer numbers, not to mention the evolutionary advantage of faster generational cycling, bacteria and viruses hold the upper hand in their relationship with humans; we have now empowered them to hopscotch across continents in hours, circling the globe by jet and spreading along trade and travel networks. As we disrupt ecosystems by changing the climate and by moving deeper into previously unexploited landscapes around the world, logging rain forest interiors and disturbing remote caves, we are also coming into contact with wild animals, and, in some cases, eating them, farming them, bringing them—and their diseases—back into our domestic lives. In the process, we are offering thousands of viruses and bacteria the chance to spill over, or jump hosts, unleashing new plagues. Quarantine is the best—sometimes the *only*—tool we have to protect ourselves against the new and truly alien. It buys us the time and space we need to respond.

Formally speaking, humans have been quarantining themselves and one another since at least the fourteenth century, in response to the Black Death. In July 1377, the maritime city of Dubrovnik, on the coast of the Adriatic Sea, instituted what are thought to be the world's first mandatory public health measures with specific provisions for quarantine. A regulation approved by the city's Major Council at the time stated that "those who come from plague infested areas shall not enter Dubrovnik or its district unless they previously spend a month on the islet of Mrkan (St. Mark) or in the town of Cavtat, for the purpose of disinfection." Rather than close the city's gates in the face of disease, sacrificing all the economic benefits of exchange, Dubrovnik's elders created a buffer, delaying the arrival of potentially infected people and goods into the city until they were proven safe.

Quarantine is thus a tool from an earlier era reclaiming center stage in our modern world, but it arrives bearing an unexpected religious pedigree: initially limited to a period of just thirty days, the eponymous *forty days* of quarantine gave the experience theological weight. By adopting the number forty, quarantine became conceptually backdated to biblical times. From then on, quarantine was a period of cleansing that explicitly referenced Christ's forty days in the desert, the forty days of Christian Lent, the forty days of rain that compelled Noah to build his ark, even the forty days that Moses spent waiting atop Mount Sinai for the Ten Commandments.

As Jane Stevens Crawshaw emphasizes in *Plague Hospitals: Public Health for the City in Early Modern Venice*, "The religious significance of the period for quarantine was not simply coincidental—it was chosen in order to bring comfort to those in need and to encourage those undergoing quarantine to look on it as a period of purification to be spent in devotion." This makes quarantine as much faith-based as it is medical, as much a purging of the soul as it is a cleansing of the body. To undergo quarantine is to distance oneself from the world in order to experience purification—then to reenter that world reborn, guaranteed free from its polluting contagion.

On a quiet, sunlit evening in September 2016, years before a novel coronavirus would shut down the world, Luigi Bertinato welcomed us into the library of the Fondazione Querini Stampalia in Venice. Bertinato, a tanned and youthful sixtysomething whose artfully rumpled hair betrayed only the tiniest hint of gray, is a medical doctor and public health policy expert. The former director of international health for Italy's Veneto region, he subsequently served as chief scientific advisor to Italy's national COVID-19 response team. He is also something of a quarantine enthusiast, deeply versed in the practice's medical history, future applications, and unique ethical responsibilities.

According to the wishes of its founder, the last descendant of the aristocratic Querini Stampalia family, the library where we met was to stay open on holidays and late into the evening—closing its doors to the public only at midnight—in order to ensure that Venetians still had a place to "study worthwhile disciplines" when similar institutions in the city had closed. (Today, the library shuts its doors at a more modest 7:00 p.m., meaning that our visit took place largely after-hours, which would lead to a moment of humor later that evening.)

Bertinato met us inside along with the historian and Querini Stampalia librarian Angela Munari, for whom he would serve as interpreter. Munari, tiny and businesslike in her archival white gloves, steered us past antique bookshelves and modern offices into a room filled with manuscripts. Some were nearly six hundred years old; all, in some way, addressed the topics of medicine, epidemic disease, and quarantine. Their often mottled and stained pages—some vellum, some paper—had been carefully laid open by Munari to display maps, official Venetian public health orders, somewhat gruesome anatomical diagrams, disinfection protocols for private residences, and illustrated recipes of putative cures for obscure ailments. The room had the rich, bodily smell of leather bindings and aged parchment, cut through by a citronella note from Bertinato's cologne, a crisp citrus fragrance with the subtlest resemblance to bug spray—a fitting coincidence, we later joked, given the city's malarial setting and struggles with mosquito-borne disease.

Together, Bertinato and Munari would prove to be ideal guides with whom to embark on our journey into the history and future of quarantine, and Venice the perfect city for this initial conversation. After all, quarantine may have first been mandated in Dubrovnik, but it was here, in Venice, that it was refined into an architectural and spatial science. Venice itself—a labyrinth of islands, canals, bridges, and docks—is a natural laboratory for experimenting with new forms of geographic control. As Jane Stevens Crawshaw writes, the city itself became an armature against the spread of disease—what she calls the "manipulation of urban space as a form of protection."

Even the Querini Stampalia sits on an island within an island, accessible only by a small, easily blocked footbridge: the very topography of the city facilitates both quarantine and isolation. The Venetian lagoon still hosts the remains of three impressive *lazzaretti*, each located on its own island, each island located progressively farther from the city center.

As we discussed this, looking at maps of Venice from different centuries open on the table in front of us and observing how prominently the city's quarantine islands were marked, Bertinato explained that a *lazzaretto* is simply a quarantine hospital. According to historians, the word—spelled *lazaretto* in its Anglicized form—most likely comes from a corruption of Santa Maria di Nazareth, the original name of the island on which Venetian city leaders built the world's first such permanent facility. The slippage from *Nazaretto* to *lazaretto* is understandable, given the fact that Lazarus—the biblical beggar covered in sores who, in the Gospel according to Luke, begs outside the gates of a rich man—is also the patron saint of lepers. This entwined etymology—a facility named either after the Holy Mother, virginal and untouched by sin, or after a diseased outsider redeemed by the grace of God—is, in a way, rather poetic: the quarantine hospital is a place for splitting the difference between purity and danger, even in its own name.

Historians believe that, prior to the 1300s, most of Europe and Asia had enjoyed centuries of relative freedom from epidemic disease. There is less agreement on where, exactly, the Black Death originated, but many believe it first flared up in or near China—a hypothesis made more compelling by reports of a mysterious sickness that wiped out a third of the Chinese population between 1330 and 1350, followed swiftly by the fall of the ruling Mongol Yuan lineage and its replacement by the Ming dynasty. What is known is that the plague had made its way to the Black Sea by 1346: historians point to a pivotal transmission event at the Siege of Caffa (now Feodosiya), a major port on the Crimean Peninsula founded by the Genoese in order to trade with the East. According to a contemporary account, the attacking Mongol forces used catapults to hurl plague-ridden corpses over the walls, which piled up in "mountains" in the streets, transmitting infection to Italian merchants and sailors who, as they fled, carried the disease back to the Mediterranean, "as if they had brought evil spirits with them."

Whether or not this particularly gruesome use of bioweapons was to blame, the plague reached Europe's ports by 1347, on merchant ships trading goods from Asia. From the start, fear of the disease competed with the enormous wealth to be earned in the spice trade. (The connection between the two was so strong that in 1348 an official at the plague-stricken papal court in Avignon wrote that "no kinds of spices are eaten or handled, unless they have been in stock for a year.")

Venice in the fourteenth century was thus a city under siege: a mysterious, highly contagious disease had begun to infect people throughout Mediterranean Europe, imperiling both its inhabitants and their source of income, and it was not at all clear how to stop it. By the end of the pandemic, Bertinato pointed out, two-thirds of the Venetian population had been killed.

Known as the Black Death, because one of its characteristic symptoms was a charcoal-like blackening of victims' extremities, the disease also caused grotesque enlarged lymph nodes in the groins and armpits of its victims. These swollen and tumescent forms are called *buboes*—thus our current name for the illness, *bubonic plague*. Piercing a bubo was horrific: doctors at the time carried a long, bladed staff for this purpose, allowing them to slice open someone's swollen glands while maintaining their distance from the reeking, infectious pus that poured forth. The sinister beaked mask of the infamous plague doctor costume—which, Munari told us, was originally designed in France but enthusiastically adopted in Venice, where it was incorporated into the commedia dell'arte and often worn during carnival celebrations—was stuffed with garlic and fragrant herbs in order to help neutralize the odor of putrescence, glandular leakage, and death. (The garlic, if freshly chopped, might actually have provided some protection: allicin, the volatile chemical that gives crushed garlic its characteristic odor, is also an inhalable antibiotic.)

To answer the question of why the formal practice of quarantine, and its accompanying infrastructure, emerged in Europe rather than in China or the Levant—regions where the Black Death had struck earlier, but with similar force—Munari explained that we needed to understand the era's disparate frameworks for diagnosing sickness and health. When the Black Death first hit Venice in the 1300s, the ideas of a second-century Greco-Roman physician named Galen still held sway in European medical discourse. Galen promoted a hypothesis that "humors"—bodily fluids including blood, bile, and phlegm, but also urine—were key to understanding human health and physiology.

Building on—and sometimes disagreeing with—the ideas of Galen were those of Ibn Sina, an influential eleventh-century Islamic physician known in Europe as Avicenna. In his book *The Canon of Medicine*, Avicenna attributed the causes of outbreaks to a complex blend of humors, miasma theory, and cosmic influences, in addition to respiratory or water-borne transmission. In such a system, an individual's lifestyle and zodiacal charts could be as, if not more, important than their exposure to bad air.

Over and above such theories, disease was almost always understood at the time through a theological framework. Across the world's great religious traditions, being stricken by plague was often attributed to divine will or cosmological forces; differences, however, arose in the suggested response. Perhaps, Venetians wondered, the plague was divine punishment for the fortune their city had built on commerce—after all, the Bible promulgates a relatively low opinion of merchants, who, it was felt, valued Mammon over God and routinely engaged in deceit. Meanwhile, according to the medical historian Mark Harrison, Islamic scholars were more likely to regard death from plague "as divine mercy or martyrdom, rather than punishment." Indeed, despite Avicenna's tentative steps toward a theory of contagion, Harrison points out that the very idea that illness might spread independently of God's will "was anathema to many Muslims." Similarly, in China, epidemic disease was historically understood as the consequence of cosmological disharmony—a situation that called for propitiating the gods rather than for constructing lazarettos.

For a society without a modern scientific understanding of infection, precisely what caused the Black Death inspired no end of speculation. In the midst of this debate, some voices began to argue that this sickness was caused by something physical, something real in our everyday world. To careful observers, the spread of the plague seemed closely connected to the arrival of people and cargo from foreign ports, places where this mysterious sickness was already known to be spreading. If a disease could come to Dubrovnik or Venice from a specific place in a predictable way, then this would make it a *secular* contagion, not a spiritual malady. The conceptual origins of quarantine were taking shape—although its rationale would be debated for centuries to come. (It wasn't until the mid-1800s, when Robert Koch made the discoveries that led Louis Pasteur to describe the germ theory of disease, that the microbial mechanism of infection transmission was finally proved.)

Keeping people and goods in a state of isolation, away from the city proper, allowed authorities to use urban form itself as a tool for field-testing this emerging medical hypothesis. Quarantine was a way to reveal a perceived connection between physical interaction, spatial proximity, and disease—proving that the plague might be best understood through an epidemiological lens, rather than an astrological or humoral one. Seen this way, Bertinato explained, the acceptance and implementation of quarantine is also an early historical bellwether for the modernization of medical practice, signaling at least a tentative trust in secular and scientific—rather than religious and supernatural—explanations for ill health.

It is no coincidence that quarantine was adopted as official policy in wealthy Mediterranean republics such as Venice, where independent citizens elected a leader, as opposed to a hereditary monarchy or feudal state. A commitment to civic virtue and an investment in the common good, embodied in the construction of a purpose-built lazaretto—or in the gift of a library such as the Querini Stampalia to the city—require a strong sense of community and shared identity. Isolation and quarantine are thus early examples of political modernity and medical rationality, of both public spirit and evidence-based science.

This is not without irony: today, quarantine is often dismissed as medieval, even primitive, yet in the time of the Black Death it was, in many ways, a remarkably sophisticated and modern act. Still, then as now, it was not without its critics. Here, we turned back to the maps and treatises that Bertinato and Munari had gathered for us, gorgeous swirls of black and red ink illustrating mountains and cities, organs and circulation. Competing medical schools in different Italian cities had different explanations for the plague, Munari said.

Nodding as he translated for her, Bertinato added that this kind of medical factionalism is a perennial problem. When the World Health Organization issued its latest revision to the International Health Regulations—the legal framework that governs the global effort to contain infectious disease threats—Bertinato was a member of the team representing Italy. He described disagreements emerging between experts from different regions of Europe, let alone distant parts of the world, holding up negotiations for hours. "The debate around those health regulations was exactly similar to the debate in Venice more than five hundred years ago," Bertinato said, articulating a recurring theme in his approach to the history and future of quarantine. We are, in Bertinato's eyes, seemingly doomed to repeat our mistakes—but also blessed to rediscover our successes.

Even in Renaissance Venice, he pointed out, practices that remain central to twenty-first-century public health coexisted alongside earlier medical theories and more superstitious notions. Munari drew our attention to a variety of medical manuscripts laid out on the table, including one from the fifteenth century arguing against the theory of contagion. What this showed was that, after the bubonic plague had burned through Europe for nearly a century, there was still no consensus on what caused it or how to prevent its spread. Across the continent, other responses to the horror of epidemic disease had proved to be both ineffective and brutal—from ancient medical techniques, such as controlled bleeding, to acts of genocidal terror, such as slaughtering a city's entire Jewish population in an attempt to appease the wrath of God. (In medieval Christian theology, not only had the Jewish people rejected Christ as the Messiah; they were also held directly responsible for his crucifixion.) Indeed, in 1516, Venice also created Europe's first official "ghetto," applying the sanitary framework of isolation to the city's Jews by forcibly relocating them to a small island in the Cannaregio district, accessible by two gated footbridges that were locked each night.

For Bertinato, the story of quarantine's adoption in Venice was thus not a triumphant tale of scientific progress over the foggy pull of superstition and scapegoating but rather a more cautionary reminder that, today, we face many of the same challenges. People are still often distrustful of medical authorities. Arguments between city officials and religious leaders, between businesspeople and public health experts, or between doctors and patients about the best way to combat the spread of infectious disease have not gone away. Hucksters and presidents alike champion unproven treatments, even as outsiders and minorities are still irrationally blamed for outbreaks. As if anticipating his own experience years later during Italy's grueling outbreak of coronavirus in 2020, Bertinato urged us to remember just how much trust, leadership, and community cohesion is required to convince people that public health measures such as quarantine, economic shutdowns, or protective face masks are in their best interest. Precisely because quarantine admits uncertainty, many people now interpret its implementation as proof that our experts and leaders simply do not know what they are doing. Ironically, given its origins, it is *quarantine* that now resembles superstition.

Maps and manuscripts came and went; anatomical diagrams followed floor plans of old quarantine hospitals, which were then replaced by maps of cruise-ship itineraries and airline routes, as Bertinato warned that cities such as Venice are still hyperconnected to a contagious world. Ebola, SARS, MERS—the outlines of a pandemic such as COVID-19 were already clear, in 2016. We ended up speaking for hours, straight through sunset—so long, in fact, that Munari eventually said goodnight, returning these precious materials to their proper locations in the archive and leaving us with Bertinato, discussing disease in a deserted palazzo.

At that point, Bertinato's eyes lit up as he revealed a surprise. He had commissioned a costume designer in the Italian theatrical world—Elisa Cobello, the daughter of a friend of his—to produce a special outfit. Reaching into his bag, he pulled out two ensembles. One was a twenty-first-century kit of Tyvek personal protective equipment, identical to what he himself had worn while in Africa treating Ebola patients; the other, sewn to Bertinato's specifications, was a plague doctor's gown from the era of the Black Death, complete with a beaked helmet that, incredibly, had been hidden inside his bag the entire time we spoke.

For the next twenty minutes, Bertinato put on his medieval Venetian plague-doctor gear, while instructing Geoff in how to don the contemporary PPE in the proper sequence. Both outfits required following a laborious, step-by-step process that was not easy to get right, even in these nonurgent circumstances. Bertinato's point was not that our present-day medical gear is as ineffectual as a beak filled with potpourri—far from it. Rather, it was that one of the primary ways that humans have responded to something we do not fully understand—whether that be the Black Death, Ebola, or, several years after our meeting, COVID-19—is to develop protocols and procedures, ritual behaviors whose performance reassures us as we encounter and attempt to

While in Venice, we met with the librarian Angela Munari and Dr. Luigi Bertinato at the Fondazione Querini Stampalia. Here, Geoff Manaugh and Dr. Bertinato don plague gear from different eras: Geoff in twenty-first-century personal protective equipment and Dr. Bertinato in the costume of a Black Death–era physician. (*Photograph by Nicola Twilley*)

triumph over the unknown. From scented herbs to hand sanitizer, and from Venetian lazarettos to high-level containment facilities, these measures curtail contact and constrain exposure, keeping an unfamiliar threat at bay.

Now fully decked out and standing beside each other—Geoff sweating inside a hooded Tyvek suit, N95 respiratory mask, and goggles; Bertinato all but invisible within his horror-movie costume—we heard a startled cough from outside in the corridor. It was a security guard. He looked utterly baffled—and more than a little frightened—by the two figures gazing back at him. It turned out that, in our fastidious efforts to ensure that we followed the correct procedure for donning protective gear, we had all lost track of time. The library had been closed for nearly half an hour. The guard, thinking he would be patrolling an empty building, had instead stumbled upon what looked like a Renaissance occult ritual crossed with a biohazard event.

Bertinato pulled off his mask and the guard relaxed, seeing Bertinato's grin. The two men said something to each other in Italian and began to laugh.

⟷

In his book on the history of monsters, Stephen T. Asma writes that humans have always lived in fear of inhuman things—creatures, forces, or influences, but also pestilences and plagues—that lie beyond the edges of the civilized world. In myths and folklore, let alone in modern horror stories, these monsters rarely remain confined to their separate realm: either we encounter them, often unwittingly, as we leave our homes and cities to travel into the unknown, or they come to us, invading our safe world from beyond.

As the classicist Debbie Felton told us, situations in which invisible dangers lurk within and seemingly harmless strangers are revealed to be monsters make up the foundation of much of ancient Greek and Roman literature. Today's horror stories about a deadly pathogen arriving at London's Heathrow or New York City's John F. Kennedy airports, transported from a remote wilderness to the heart of our civilization overnight, find an ancient echo in the expanding road networks of classical Greece and Rome. New transportation infrastructure brought previously isolated city-states into proximity, Felton said, which facilitated commerce—but also triggered enormous anxiety, expressed in an explosion of highway-based monster stories.

"These stories about monstrous encounters on the road are really about contact with unknown peoples," Felton explained. "There's just a natural uncertainty and fear about the wisdom of this kind of interaction."

Asma goes on to suggest that what it means to be human is often defined by our degree of separation from what we think we are not. As part of the rich cycle of myths that developed after the death of Alexander the Great, one story of isolation stands out. "Alexander supposedly chased his foreign enemies through a mountain pass in the Caucasus region and then enclosed them behind unbreachable iron gates," Asma writes. Beyond this wall was what Asma calls a "monster zone"—a "prison territory"—in which Gog and Magog, enigmatic creatures that play an almost Godzilla-like role in the biblical book of Revelation, were locked away in a realm of grotesquerie and violence. The Caucasus Mountains being the supposed site of origin of the Caucasian people—a pseudoscientific term used by white, Enlightenment-era Europeans to classify themselves in opposition to other races—it is striking that the story of these gates suggests that the very idea of the West required an act of isolation against a monstrous other that lurked somewhere in Eastern darkness. The story of Alexander's gates, Asma writes, was so widely told and retold in medieval times that world maps of the era often feature this Caucasian barrier as if it were a confirmed geographic landmark.

Nevertheless, as with all legends of isolation and quarantine, the separation cannot stand: the monsters behind the wall are destined to break free, their incarceration only temporary. There is always a weakness, a point of vulnerability, a crack in the masonry. As the literary theorist Jeffrey Jerome Cohen has written in an essay on the role of monsters in literature, folklore, and mythology, "The monster always escapes"—indeed, Cohen adds, "The monster's very existence is a rebuke to boundary and enclosure."

In this case, a curious fox is to blame. Asma summarizes a fourteenth-century work called *The Travels of Sir John Mandeville*, explaining that "during the time of the Antichrist a fox will dig a hole through Alexander's gates and emerge inside the monster zone. The monsters will be amazed to see the fox, as such creatures do not live there locally, and they will follow it until it reveals its narrow passageway through the gates. The cursed sons of Cain will finally burst forth from the gates, and the realm of the reprobate will be emptied into the apocalyptic world."

Our evening spent exploring the stacks of the Querini Stampalia library reinforced just how much of Western literature consists of stories of people attempting to isolate themselves from a threatening world, as in Giovanni Boccaccio's fourteenth-century collection of plague novellas, *The Decameron*, or of people waiting in uneasy horror to see if they are becoming something they do not want to be. In this light, even werewolf stories can be stories of quarantine, as potential exposure to a monstrous affliction—lycanthropy—inevitably leads to a scene of agonized waiting for the first signs of transformation to begin. As such a comparison shows, quarantine rapidly slips from the realm of medical practice into pure metaphor.

Consider Prince Prospero and his guests in Edgar Allan Poe's classic story of medical isolation gone wrong, "The Masque of the Red Death." Poe depicts a decadent world wracked by a gruesome and fatal hemorrhagic fever that causes "profuse bleeding at the pores." Because of this, it is named the "Red Death." Fleeing for their safety, a thousand guests seal themselves inside Prince Prospero's palace, going so far as to weld shut the outer gates. "With such precautions the courtiers might bid defiance to contagion," Poe writes. "The external world could take care of itself."

Thinking they have locked themselves away from pestilence, the guests instead find they have locked themselves in with the infection. A costumed figure steps forth from the crowd bearing obvious symptoms of the disease, but, with no way to escape, "one by one dropped the revellers in the blood-bedewed halls of their revel, and died each in the despairing posture of his fall." At the end of Poe's fable, everyone in the palace has died, the flames of a thousand candles and chandeliers have burned out, "and Darkness and Decay and the Red Death held illimitable dominion over all."

We are always just one breach of the barrier, one failure of containment, away from devastation, these stories claim. It is these fears—of a pulsing, infectious, contaminating, monstrous other—that so animate the topic of quarantine and give it its chilling, perennial fascination. It seems telling that a 2008 movie about the outbreak of a genetically engineered strain of rabies in a Los Angeles apartment building was titled simply *Quarantine*, as if the word itself needed no further explanation to inspire horror.

⟷

For a tool so potent, quarantine has surprisingly few rules, yet its apparent lack of limitation is exactly what gives it such power and flexibility. As our long evening with Luigi Bertinato came to a close—his plague gown and Tyvek suit neatly folded up and put away, the three of us about to part ways and head out into the Venetian night—he wanted to make sure we understood the basic guidelines that define when and where quarantine takes place.

The most important rule of all—one that has defined quarantine since its inception and still underpins WHO and CDC guidelines today—is that quarantine *requires uncertainty.* In other words, if you know that you are infected with a communicable disease, and if you have been told to stay at home or in a hospital to avoid spreading that disease, then you have not been quarantined: you have been *isolated.* This means that, by definition, quarantine emerges from a state of suspicion: it is about *potential* infection and *possible* risk. When, in October 2020, President Trump was diagnosed with COVID-19 and tweeted that he was beginning a "quarantine and recovery process," he was, of course, misusing the term: he was going into isolation.

Over the many years we spent researching this book, curious friends often asked us about leper colonies, tuberculosis (TB) sanatoriums, or the infamous case of Typhoid Mary, who was held against her will by public health authorities on a remote island in New York City for nearly three decades. But all of these are, in fact, instances of *isolation* not quarantine. Typhoid Mary, for example, did have typhoid, and, although her treatment by medical professionals was inhumane, she was known to be infectious, although asymptomatic. Typhoid Mary is not a story of quarantine.

This also means that an oft-cited biblical injunction, from Leviticus 13:46—"All the days wherein the plague shall be in him he shall be defiled; he is unclean: he shall dwell alone; without the camp shall his habitation be"—is, technically, a call to isolate the sick, rather than an instruction to quarantine the potentially exposed. The two terms are often used interchangeably, even by medical professionals; they both involve detention in the name of public health and often appear identical in practice. Nevertheless, Bertinato stressed, they refer to different things—a difference that is medically, legally, and philosophically consequential. Quarantined individuals are—for the time being—healthy. We simply have reason to believe that they may yet become sick.

As a corollary to this, Bertinato continued, quarantine, at some point, must end. If you are in a state of permanent quarantine, then you are not, in fact, in quarantine. You have been isolated—imprisoned, even. The forty-day time-span that gives quarantine its name was and has always been arbitrary. Every true quarantine reflects the incubation period of the germ it is intended to contain. For some diseases, you need to wait mere days to determine whether you have been infected; for others it can be two weeks or more. Either way, you will, at some point, be freed from quarantine (even if only to be transferred into isolation because you have been confirmed to be infectious).

Quarantine, Bertinato warned, is also useful only for some diseases and not for others. At the very least, an illness must be communicable from one person or organism to another, otherwise there is no point in quarantine. Stemming directly from this, there is very little sense in using quarantine if we can easily test for the disease in question or be protected from it by treatment and vaccination. If we can be diagnosed quickly and accurately, then there is no uncertainty; if you have an illness that can be treated, managed, or cured outright, then quarantine has no purpose. Diseases that are transmissible before symptoms appear—such as COVID-19—are both unusual and require quarantine: without clear indicators of sickness, their diagnosis can be delayed, making uncertainty inherent to their insidious spread.

Finally, quarantine is not obsolete. The fact that it is of no use in many situations should not distract from the fact that, in the case of novel infectious diseases, quarantine is often our *only* defense. Before a cure, before a vaccine, before the transmission and characteristics of a new pathogen are understood, the only thing medical authorities can do to slow the spread of disease is to reduce the frequency, duration, and variety of our interactions with one another, using isolation and quarantine. Bertinato—along with dozens of other medical experts whom we interviewed for this book, from the CDC to WHO to DARPA—knew that an outbreak like COVID-19 was coming. In early 2018, public health officials dubbed this abstract future contagion Disease X: an epidemiological placeholder to remind planners and policy makers that unknown pathogens with pandemic potential were certain to emerge in the coming years, with ever-growing frequency.

COVID-19 was the first Disease X, but it will not be the last: Bertinato and his global colleagues expect that we will be quarantining *more* in the near future, not less. You or someone you know very likely experienced quarantine during the coronavirus pandemic, and you or someone you know will very likely experience quarantine again. You may even be in a state of quarantine as you read this. This spatial tool from another era has reemerged in a world coddled by flu shots and over-the-counter cures. Quarantine, a solution from the past, is back—and here to stay. "We need a futurologist of quarantine," Bertinato said as he departed.

Over the past six centuries, quarantine has shaped the public health response to infectious disease around the world—but it has also shaped our streets, buildings, and cities, our borders, laws, identities, and imaginations. Quarantine has inspired the construction of great fortresslike facilities, built on the edges of civilization, as well as high-tech medical institutions in the very heart of the modern metropolis. While reporting this book, we crawled into crumbling hospitals overlooking the sea, toured ruins overgrown with weeds, and donned hard hats to step inside a brand-new federal quarantine facility then under construction in the geographic center of the United States. Quarantine has also transcended its biomedical origins to become a vital tool in protecting our global food supply and even our planet; our travels took us to a greenhouse in suburban London charged with safeguarding the world's chocolate supply, to an animal-disease research center in Manhattan, Kansas, built to survive the strongest tornadoes, and to a pristine spacecraft-assembly room in Pasadena, California. Quarantine is not just the purview of the WHO and the CDC—as we discovered, officials at the U.S. Department of Agriculture and NASA also depend on it to stave off famine and safely explore the cosmos.

We realized that, by examining what, where, and why we quarantine, we were not only exploring the limits of scientific understanding but also excavating our deepest fears, biases, and sense of identity. Quarantine reveals how we define and police the perimeters of self and other, as well as what we value enough to protect—and what we are willing to sacrifice. All too often, we discovered, quarantine is flawed and leaky, even deeply unjust. Almost always it has been designed simply as a buffer, rather than thought through as a lived experience. Sometimes it is the only thing that has saved us from death and devastation.

This book is about how we got from public health restrictions on the shores of the Adriatic Sea to chainsaw-wielding mobs off the coast of Maine, from magnificent brick fortresses on marshy islands in the Venice lagoon to roadside motels painted black in Washington State. By exploring quarantine across time and space, at scales ranging from viruses too small to be seen with the human eye to the unimaginably vast distance between planets, and from ruined lazarettos to salt mines in New Mexico, we sought to shed new light on this powerful and dangerous tool—in the hope that we might use it more wisely in the future.

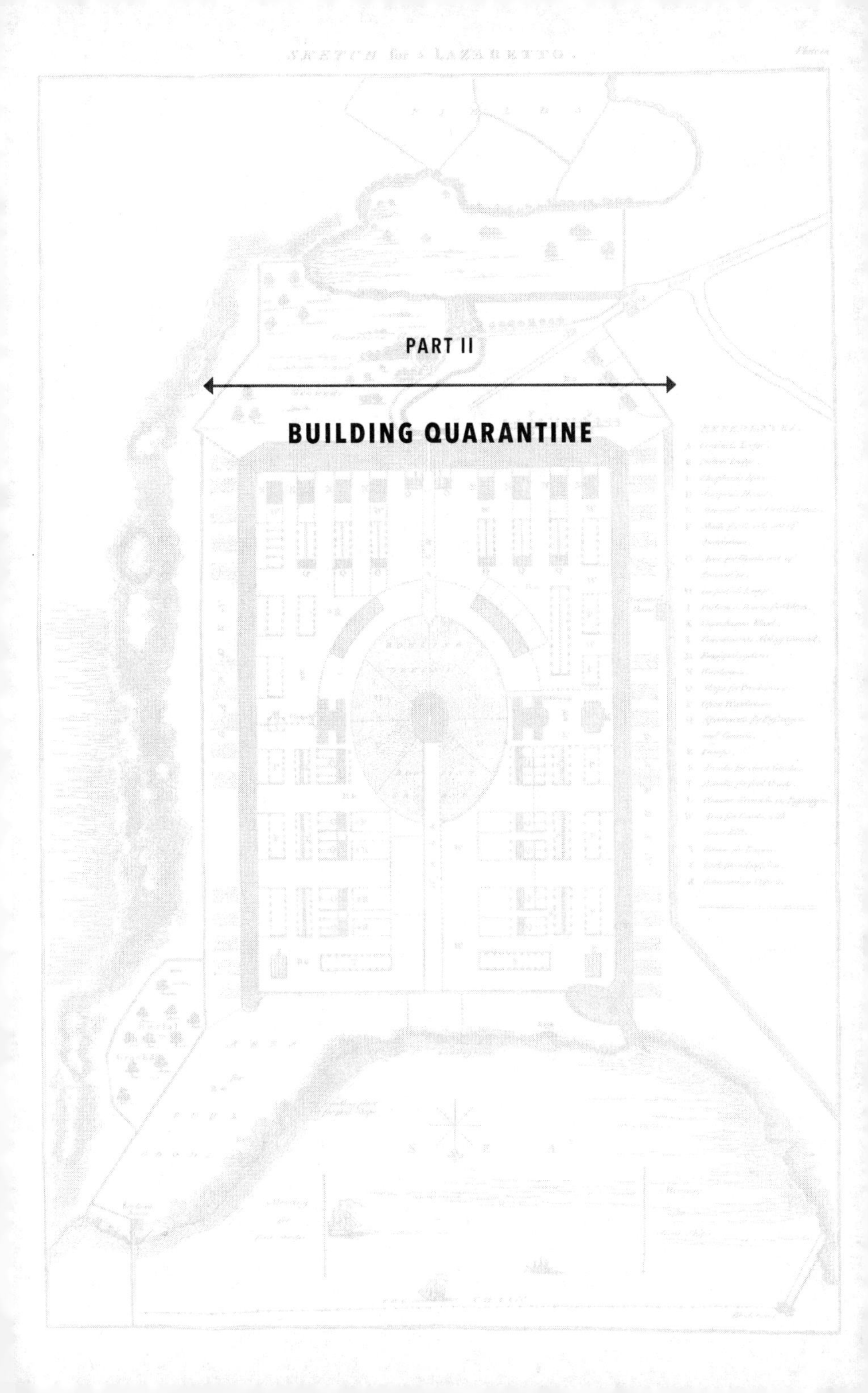

PART II

BUILDING QUARANTINE

2

The Quarantine Tourist

At the end of November 1785, the British philanthropist, vegetarian, and prison-reform advocate John Howard departed London on a trip that would take him through the Netherlands to France, heading south overland for Marseille. From there, Howard would continue onward through Nice and Genoa, then, by sea, to Malta and, beyond, to Constantinople. At that point, already in his late fifties, Howard had traveled more than forty-two thousand miles, visiting prisons, jails, and dungeons throughout Europe. Howard's goal, for more than a decade, had been to inspect the conditions experienced by people incarcerated in cities and kingdoms across the continent; many of these prisoners, held in states of hunger and isolation, had been locked away for nothing more than personal debt. Howard had become resolute in his belief that architecture—including the design of prisons—should serve humanitarian purposes, but the worst facilities he visited were more like charnel houses, he later wrote, "almost totally unprovided with the necessaries of life."

Enriched by an inheritance from his father, who had made a small fortune in the upholstery trade, Howard spent today's equivalent of nearly $3 million on his travels, an exorbitant sum that took him from Ireland to Russia, Portugal to Greece, as well as throughout the United Kingdom, in an age when such distances could be covered only by ship, foot, or carriage. It was slow going and, despite the cost, uncomfortable. At times, Howard's numerous biographies read more like adventure novels. During his journeys abroad, Howard would be forced to flee France in disguise to evade royal authorities, help to fight off a Tunisian pirate ship near the coast of Greece, and cross the Adriatic Sea on a boat whose crew had been infected with bubonic plague, returning home to testify before the House of Commons in a bid to improve prison conditions throughout the United Kingdom.

Howard, whose life's work would eventually be honored with the first statue of a nonreligious figure installed in London's St. Paul's Cathedral, is often described as "the prisoner's friend": his published writings and parliamentary evidence are credited with inspiring subsequent legislation to improve conditions in jails, and some of his proposals were implemented in the construction of new prison buildings. His legacy continues today in the U.K.'s Howard League and the John Howard Society of Canada, both of which campaign for carceral reform.

The motivation behind Howard's self-imposed mission remains obscure. His contemporaries viewed him as eccentric, obsessive, and extremely rigid in his timekeeping and habits. (One of his biographers notes that he tenderly informed his wife-to-be that "to prevent all altercations about those little matters which he had observed to be the chief grounds of uneasiness in families, he should always decide.") After dabbling in potato breeding, meteorology, and the design and construction of model cottages for workers, Howard finally found his calling at the age of forty-seven, when he was appointed High Sheriff of Bedfordshire, with responsibility for inspecting the local jail. The conditions appalled him—he was particularly horrified to find that innocent people were required to pay a fee before being released following their trial—and he set off to inspect sites of detention across the length and breadth of Europe, publishing his findings in 1777.

In the course of his travels, Howard had noticed that European nations maintained a permanent network of lazarettos and quarantine stations—a health infrastructure that, at the time, was entirely lacking in England. Such facilities—part hospital, part jail—had a reputation for being squalid and unsanitary. People whose only crime was suspected exposure to contagious disease were detained in cramped rooms "saturated with infection," as Howard would later write: windowless wards, infested with vermin, without furniture or other basic comforts.

Howard had become famous—admired and reviled in equal measure—for knocking on the gates of institutions without warning. Once inside, he would not only demand a full tour but insist on assessing every detail: counting the number of windows while weighing the inmates' bread ration with his own portable set of scales. Like the prisons Howard had been inspecting for more than a decade, Europe's quarantine facilities appeared to have no broadly recognized standards, either for their staff's medical training or for their architectural amenities.

At the same time, Britain's growing economic dependence on international trade meant that the question of quarantine was increasingly urgent. Plague was still considered an imminent threat: an outbreak in southern France had resulted in at least one hundred thousand deaths in the 1720s. (This was the last major outbreak of bubonic plague in Western Europe, but, as Howard embarked on his travels, no one could have known that.) What Howard had come to call "gaol fever"—most likely typhus, augmented by malnutrition—was rampant. For all the improvements Howard had helped to deliver for the imprisoned, no one had yet undertaken a similar project to inspect the conditions of the quarantined. This latest trip, he hoped, would change that.

⟷

As we slipped through a hole in the wall into the ruins of a five-hundred-year-old unfinished lazaretto on the island of Lokrum, just offshore from Dubrovnik, we wondered what Howard would make of our approach, which was often similarly spontaneous if slightly less metrics-obsessed. Seeing firsthand what remains of Europe's earliest quarantine stations seemed like the most sensible method for shaping our early research. We packed Howard's collected works into our hand luggage and flew to Europe, pursuing our own twenty-first-century itinerary while following in his eighteenth-century footsteps wherever we could.

Quarantine, as Howard's criticism of the practice showed, is an inherently architectural phenomenon, in the sense that it subjects people to new kinds of space and time, constraining their interaction in the name of controlling infection. By seeing how the concept has been expressed in built form, we hoped to understand both its fundamental principles and its evolution. Quarantine facilities necessarily embody the dominant medical theories of the time, their structure and flow revealing how physicians of a particular era believed diseases could spread. A sophisticated understanding of germ theory calls for very different material and ventilation choices than a medieval faith-based approach to sickness, or an eighteenth-century adherence to miasmatic theory. Lazaretto design also inevitably reflects an era's prevailing values and biases, whether through the perpetuation of class status or the provision of particular cultural amenities.

We also wanted to see *where* quarantine has historically occurred—not just the form of individual buildings, but their locations. Changes in lazaretto placement offer a guide to the shifting geography of trade and travel, while also gesturing at new health threats, both real and imagined. They also serve as a marker of a society's confidence in its technologies of containment. In medieval Dubrovnik or Venice, siting a quarantine hospital was a balancing act: it should not be so far away as to inconvenience merchants, and it had to be close enough to visibly reassure residents of its protection—yet it should also be sufficiently remote and isolated so as to offer no chance of escape. Today, biocontainment facilities are located in the heart of some of the world's largest cities; quarantine has moved from the peripheries and edges to the center.

Last, but far from least, we hoped to get a sense for how quarantine has been experienced for hundreds of years: the sights and sounds, the spatial ambience of this once—and now once again—ubiquitous form of medical detention.

Howard's travel diaries, letters to colleagues, and publications describe fortified buildings on cliffs overlooking the sea; sprawling complexes, with their own churches and cemeteries, dominating entire islands; and remote and imposing facilities like the one he would visit in Marseille, a building perched atop a rock, he wrote, that "commands the entrance of the harbour." But a twenty-first-century John Howard does not have so many places to visit. Many of the most impressive lazarettos and quarantine stations—the highlights of Howard's idiosyncratic grand tour—have since been demolished. Those buildings survive in photographs, at best, and, more often, as brief annotations in ship captains' logs or as postal cancellations and disinfection marks on letters mailed by travelers locked in quarantine. In other cases, lazarettos have crumbled into ruins, or, if they remain standing, are so thoroughly altered as to be unrecognizable, having been given new lives as hotels, art centers, storage facilities, or administrative offices.

There is at least one other reason why quarantine has left so few fossils in our contemporary world: permanent lazarettos have, historically, been something of an anomaly. The costs involved in the construction and maintenance of such facilities are, for most, overly burdensome. Far more frequently, quarantine has taken place in temporary facilities: hastily retrofitted barracks, ships, or monasteries, or quickly erected wooden huts and tents. (In almost all cases, such huts were subsequently dismantled or burned to the ground once the epidemic had subsided, as a final act of disinfection.) What most quarantine facilities of the past really looked like—let alone what they felt like to someone

The unfinished lazaretto on Lokrum, an island just offshore from Dubrovnik, Croatia, is now a partially stabilized ruin. *(Photograph by Nicola Twilley)*

on the inside, stuck there with fellow crew members, traveling companions, or strangers—is one of history's many unknowns.

Here on Lokrum, visitors would be hard-pressed to guess what this vast sandstone wall hidden in the woods, with no visible signage or even a roof, really was. After walking around three sides of its square perimeter, avoiding ankle-twisting piles of loose rocks that had spalled from the old masonry, we found a gap in the wall and slipped through. We emerged in a large, meadow-like space. The silence was broken only by the sound of passing boats and by our own footsteps rustling through fragrant vegetation. The smell, as we crushed the long stems underfoot, was herbal, slightly sunburned. We wandered around with our cameras and notebooks, pausing here and there beneath silvery groves of olive trees for some shade, and stepping over the remains of collapsed interior walls no higher than our knees. Alcoves in the masonry gave at least a partial indication of where rooms would have once stood, hinting at the genuine complexity and ambition behind the lazaretto's design. Our rambles soon attracted attention: a family of curious donkeys began walking toward us up a small slope. The animals—the lazaretto's only other inhabitants—seemed bored and lonely (or perhaps just hungry). They proceeded to follow us around, sometimes nuzzling up against us as we stopped to write notes or take a picture.

Visible across a narrow azure channel from Lokrum's quarantine ruins, Dubrovnik is the kind of European city you will have seen many times before you ever set foot there. Its meandering, castellated walls, cobblestone streets, narrow hillside stairways, and exquisitely carved, oversize doors opening onto the interiors of medieval churches would not be out of place in a fantasy novel. In fact, today, many people's first introduction to Dubrovnik comes from its use as a shooting location for HBO's series *Game of Thrones*, several scenes of which were also filmed on Lokrum.

For centuries, Dubrovnik—then known as Ragusa—served as a first stop for goods and people entering Europe from the Middle East and North Africa. A ship or caravan en route from Lebanon to Venice—its cargo destined farther inland for, say, Milan or Munich—would likely have broken its journey in Dubrovnik along the way. Ideally and picturesquely situated on the eastern shores of the Adriatic, Dubrovnik was at the front line of encounter between the European West and its exoticized Oriental other. With travel and trade came disease—and with disease came the need for quarantine.

Construction of the lazaretto on Lokrum began in the 1530s, although the building was never completed—most likely, historians have concluded, out of fear that its thick walls, designed to keep a deadly contagion inside, might instead be commandeered by would-be invaders, giving Dubrovnik's foreign enemies a fortified base of attack a mere thousand feet offshore. Despite its half-finished state, it was used to isolate the sick during a plague outbreak in the 1690s, before being left in peace for the next three centuries, slowly falling into ruin.

Even without a planned quarantine facility on Lokrum, however, the republic of Dubrovnik was able to rely on an expansive network of outlying islands for medical protection. These included an island-within-an-inland—St. Mary's, a tiny gem tucked in the cove of a sparkling turquoise saltwater lake on the island of Mljet. The Benedictine monks of St. Mary's were herbalists—in effect, medical gardeners—and their doubly isolated monastery served the nearby city as a quarantine facility for more than a century, from at least as early as January 1397 until 1527.

But the separation provided by a watery cordon, on its

own, was not enough: an effective quarantine required enforcement. In Dubrovnik, this meant sending armed sailboats out on regular patrols to deter those held in the city's lazarettos from swimming ashore, as well as to prevent unauthorized landings by ship captains hoping to bypass quarantine altogether.

As the historians Zlata Blažina Tomić and Vesna Blažina have discovered by digging through the city's archives, most tales from Dubrovnik's early experiences with quarantine involve people trying to evade the city's health regulations. Their colleague Vesna Miović recounts the tragic tale of Dominko, a boy of just thirteen or fourteen, who was beaten so frequently by his parents that he slipped into the lazaretto in search of a safe hiding place, and a merchant allowed him to stay. Dominko was discovered when soldiers overheard the two conversing; because he had potentially been exposed to disease, he was ordered into quarantine in a different lazaretto. "We do not know what happened to him next," writes Miović.

Much more common than attempts to break into quarantine were attempts to smuggle goods out of it—or steal from the quarantined. During major epidemics, such as that of 1526–27, when a quarter of Dubrovnik's population died, the nobility fled the city, while most commoners remained. (According to Tomić and Blažina, Dubrovnik officials ruled that "only those who had no specific tasks to perform during a plague outbreak could leave"—in other words, the medieval equivalent of essential workers were required to stay, in order to keep the city running.) Public health authorities then moved any potentially sick people to the lazaretto, leaving even fewer residents behind. The resulting empty homes frequently proved too tempting for the city's struggling poor.

People who had already recovered from exposure to the

plague, giving them at least partial immunity, made excellent burglars. In Dubrovnik, organized criminal groups began a brisk business breaking into the homes of the quarantined, which had helpfully been marked for them by health authorities as a warning to stay away. Ambitious burglars even began to target quarantine stations, breaking in to steal valuable merchandise that had been locked up specifically to protect the public. These sorts of tainted objects could prove difficult to track and control, carrying invisible contagion into the city.

Public health officials often relied on plague survivors—women were hired to clean and disinfect homes and merchandise, while men worked as gravediggers—but they also mistrusted them. The first people to be sentenced to death for breaking the city's health-control measures were two gravediggers who were caught stealing from a lazaretto, and publicly executed. "It goes without saying that the social constellation in Dubrovnik was such that not a single patrician was ever hanged for violating plague control regulations," add Tomić and Blažina.

These early enforcement challenges have proved endemic to quarantine over the ensuing centuries: the historian Jane Stevens Crawshaw describes an anonymous complaint from Genoa in 1656, claiming that the body cleaners and laundresses were "engaging in sexual commerce" and "passing goods out illicitly to strangers." Centuries later, the city of Melbourne, Australia, traced much of its second wave of COVID-19 cases to "breaches of protocol" by low-wage private security guards charged with enforcing the country's quarantine program for travelers—breaches that allegedly included having sex with individuals under their supervision.

Quarantine powers meant that the capacity to incarcerate and confine people in prisons had been fused with the power to determine health and illness. Dubrovnik's new laws meant that the city could preemptively remove people from society who *might* be sick or infectious—but they also helped to stoke suspicion and closer government scrutiny of already-marginalized people. Laundresses, gravediggers, and guards were doubly hard-hit: they were exposed to disease by virtue of their jobs, but also singled out politically, to be treated as suspected carriers of plague whose everyday movements now needed to be managed and controlled.

According to Tomić and Blažina, Dubrovnik was not only "the first government in the world to formulate, develop, and apply the concept of quarantine"; it was also the first to establish a health office, and to make it into a permanent institution of city management. Defending its subjects from medical threats was now part of the republic's long-term mission. "Controlling the space and the movement of the few in the interest of the common good" had become an essential responsibility of enlightened government, as Tomić and Blažina put it.

Central to achieving medical security was the use of widespread data collection and epidemic intelligence. Dubrovnik's officials were pioneers in the aggressive gathering of news about outbreaks anywhere in the region, from Europe to the Middle East and North Africa, but particularly in lands bordering the Adriatic Sea. Boat by boat, house by house, information about potentially exposed family members—as well as foreign merchandise of unknown origin—was subject to official collection. Seafaring merchants were asked about other vessels they encountered at sea, compiling a catalog of possible maritime spread. This was an early and rudimentary attempt at what we might call *contact tracing*, and one of the best places to start was at the quarantine station itself. "The officers of this Lazaretto serve and accommodate travelers and learn many of their secrets and private affairs," wrote the Ottoman explorer Evliya Çelebi, describing quarantine at Dubrovnik in his travelogue. According to the historian Vesna Miović, the city received reports on everything from illegal smuggling of foreign merchandise to political turmoil in Bosnia. "It was easy to find out all the news in the Lazaretto," she writes.

Today, Dubrovnik's last and best-preserved lazaretto is still buzzing: it has been renovated and reinvented as a creative complex. Just outside the Ploče gate, within walking distance of the small apartment we had rented within the city's outer walls, is a tidy seaside complex of small buildings, each featuring arcade-like arches that now house a theatrical space, art galleries, and several open-air courtyards. The Kavana Lazareti, a restaurant with a delightful outdoor terrace, gives away the facility's original purpose in its name.

The Dubrovnik Senate ordered the construction of a new quarantine station in 1590, in response to a shift in trade patterns: more merchants from Constantinople were arriving by road, and this site, at the city's eastern gate, was perfectly placed to intercept them. Construction finished around 1647. The resulting lazaretto followed the era's by then standardized template for the architecture of quarantine. The complex was entirely enclosed behind high, almost windowless walls; its two locked gates, one opening out to sea and the other on the land side, were monitored by guards in the lazaretto's twin towers. Inside were nine separate spaces used for isolating travelers on a ship-by-ship basis; alongside those were five other holding rooms reserved for goods and cargo. A series of outdoor courtyards allowed for sunlight, ample breezes, and natural ventilation, which helped, it was believed, to dissipate the bad airs of potential infection. Goods were disinfected in the courtyard arcades, while merchants were housed in the story above, in spacious rooms with hearths and access to a terrace, complete with stone bench.

The lazaretto at Ploče was the pride of Dubrovnik's anti-epidemic system: new introductions of the plague fell after it was opened, a relief for which the facility was given substantial credit. No one knows exactly when it ceased to operate. With superb views, an ideal location on the city's rocky shore, and a series of well-shielded and spacious courtyards capable of hosting multiple simultaneous performances, the lazaretto is now a rather wonderful place to spend time. We came back more than once, wandering around art galleries at the end of a long day, enjoying a drink together outdoors, and reading back through our notes and research materials.

One night, we walked by to see that a program of Croatian folk dances was about to begin inside one of the performance spaces, and we bought tickets. The event, we immediately realized, was targeted at elderly cruise-ship passengers, but we were nevertheless dazzled by the juxtaposition of lively songs and choreography with the site's history. It was somewhat jarring to realize that the spatial conditions that once made the lazaretto at Ploče such a successful quarantine facility are, it seems, the same ones that make it a vibrant arts and culture center today.

⟷

As the coast of Croatia disappeared in the distance, we stood on the deck of our ferry, gazing back at the huge walls of Diocletian's Palace before they disappeared from view, obscured by bright lights and commercial signage. We had traveled by sea from Dubrovnik to Split, winding through the region's labyrinth of islands, including Mljet, with its Benedictine monastery, and were now en route west across the Adriatic. A dozen other travelers had come out onto the deck with us as the sun went down. Some leaned over the rails, taking photographs of each other, the ship's wake spreading slowly behind them across a quiet sea. A few minutes later, the Moon crept up over the horizon and, finally, all land disappeared.

Earlier that day, we had met with the architect Snježana Perojević in her office in the northwest tower of the old Roman palace to learn more about Split's brief but significant history with quarantine. Dressed in industry-standard black, Perojević opened her laptop and projected images of the city's long-demolished lazaretto onto the massive white limestone walls. Diocletian's Palace, built in the fourth century as a retirement home for the Roman emperor, still shapes the urban core of Split today; the shops, restaurants, homes, and workspaces of the modern city are woven seamlessly through the remains of the 1,700-year-old fortress. The palace is the structure that gives form to Split—but the lazaretto, Perojević told us, although almost no trace of it remains, is the building that ensured the city's survival to this day. "I always say we must know about the lazaretto like we know about the Palace of Diocletian," she said. "Why? Because it saved Split."

Using drawings, records, and a few pre-demolition photos, Perojević has virtually reconstructed the enormous facility, which, at one point, was one-fifth the size of the entire city. Prior to the lazaretto's construction in 1582, Split was a fairly modest town of no great economic importance: trade from the Levant traveled by sea to Venice, stopping at Cyprus, Crete, or perhaps Dubrovnik. The journey was long and dangerous—pirates prowled the Mediterranean, and, as Perojević told us, "It is a very wild sea." A Venetian merchant, Daniel Rodriga, suggested a workaround: trade caravans could travel overland through Ottoman-controlled territory all the way to Split, at which point their final destination would be just a short boat ride away, across the much calmer and heavily patrolled Adriatic.

Rodriga put his own money into the initial construction of the lazaretto at Split; the Turks built bridges, roads, mosques, and rest areas for caravans on the forty-three-day journey from Constantinople; and the Republic of Venice provided the remainder of the funding necessary to operate and expand the facility. From a financial point of view, the enterprise was a huge success: Perojević calculates that every single year between 1588 and 1641, the Republic of Venice made back double the amount it originally invested in the lazaretto's construction.

By the mid-1600s, relations between the Ottoman Empire and the Venetian Republic had become unstable. When, in 1648, Ottoman forces embarked on a lengthy siege of the Venetian-controlled town of Candia (now Heraklion) in Crete, Split's medieval walls suddenly seemed insufficient—but military experts agreed that the city's topography and geographic situation made it impossible to defend effectively. In documents that Perojević uncovered, an Italian army engineer named Innocento Conti wrote that Split's position was such that "it can't be fortified in the proper way." The only thing to do, Conti concluded, was to move the city's inhabitants to offshore islands and to demolish the entire town so that the Turks couldn't occupy it. "If they followed his advice," said Perojević, "today we wouldn't have Split."

Fortunately, they did not: the Republic of Venice felt it simply could not afford to lose Split's lazaretto. In the archives, Perojević discovered a report from Split's Venetian governor-general in which he wrote that the money that flowed from trade through the lazaretto was so great that Split had to be protected at any cost. "It is very nice, if you read it in Italian," Perojević told us. "He said: this money is like the very nerves, blood, and soul of the Venetian state."

In the end, Venice paid for the construction of three forts, and Split was neither invaded nor demolished. Its lazaretto, however, never really bounced back from the midcentury Ottoman-Venetian hostilities: Dubrovnik's brand-new Ploče gate lazaretto seized much of the caravan trade, and the Split facility shrank. One section was torn down to build a railway; other sections were repurposed as a theater, a warehouse, and even, during the Croatian Fascist regime in the 1930s, a prison and torture chamber. When the structure was severely damaged by Allied bombing during World War II, the city decided to demolish it altogether.

"It wasn't important as architecture," Perojević said. "They were very simple buildings with no decoration." There seems to be no record of who originally designed Split's lazaretto; Perojević's research reveals that the complex used the same floor plan as Dubrovnik's later facility. "Every lazaretto looks basically like a prison, because, in a way, it is a prison," she said. Ultimately, Perojević told us, although the lazaretto saved Split, it wasn't much loved. The economic benefit accrued almost entirely to the Venetian republic, while the inhabitants of Split bore all the risk. "Really, it was dangerous for Split—a few epidemics came because of the trade," she said.

Perojević sent us on our way with a story illustrating the ease with which quarantine can fail. In 1784, she told us, an official working at the Split lazaretto became enamored of some imported fabrics that were being held in quarantine. Ignoring the danger, he smuggled home a beautiful white scarf for his wife. Lodged inside its dense weave, however, were fleas carrying the *Yersinia pestis* bacteria—the cause of bubonic plague—and, with his stolen gift, the man gave his wife the Black Death. In his quarantine research, John Howard mentions the subsequent outbreak, noting that it killed one in ten of the city's residents.

For his own crossing of the Adriatic Sea, a couple of centuries before our own, John Howard had become fixated on the idea that, despite years of firsthand inspections, interviews with lazaretto staff, and a growing collection of sketches and floor plans, he had seen Europe's lazarettos only through the eyes of an outsider, a mere tourist. Howard worried that he lacked any understanding of what it was really like to undergo quarantine, to be subject to its unusual constraints and limitations. He decided that the best way to improve the accuracy of his reporting—to become a better critic of quarantine—was to experience a lazaretto in the same way so many tens of thousands of unfortunate others had done before him.

As Howard later wrote, "on farther consideration I determined to seek an opportunity of performing quarantine *myself.*" To achieve this goal, he would have to arrive at his next destination on an infected ship. This revelation came to Howard while he was in Constantinople, from whence he promptly traveled to Smyrna—present-day İzmir, on the west coast of Turkey—and, having found "a ship with a *foul* bill," set sail for Venice. The resulting journey would take nearly two full months at sea—two months during which he risked contracting an incurable and deadly illness at any moment. "I go where none of my conductors have courage to accompany me," he wrote to a colleague in a letter posted just before leaving Constantinople.

While the circumstances of our own ocean voyage were much less dramatic—we were not, to our knowledge, aboard a diseased vessel or destined to spend any time in quarantine, but rather confined to tiny, separate bunks, each of us wearing earplugs against the ship's deafening engine roar—it was nevertheless a thrill to arrive in Ancona, on the east coast of Italy, by sea the next morning. Our ship navigated the harbor waters at sunrise as we stood outside, marveling at the great, geometric walls of the city's lazaretto.

Designed by Luigi Vanvitelli, one of the most celebrated Italian architects of the eighteenth century, the lazaretto at Ancona is an enormous brick pentagon built on its own artificial island in the port. Our guide to this, one of the most awe-inspiring historic quarantine facilities to have survived, was Fausto Pugnaloni. Pugnaloni is an architect, historian, and coauthor of a beautifully illustrated book documenting the Ancona lazaretto.

"This was a completely original design," he told us as we embarked on a lengthy tour of the complex, walking its outer dock walls as well as its tunnel-like interior corridors. The building's unusual form, he said, had been chosen for its symbolism, referencing the human body. As in Leonardo da Vinci's drawing of the so-called Vitruvian Man, the head, arms, and legs each supply one point of the pentagon. Its design is also reminiscent of a star fort: the pentagonal form that military engineers relied on to defend against cannon fire had been repurposed to provide at least symbolic protection against an invisible enemy.

Pugnaloni gestured at the three-story walls. "The Adriatic is a highway, and this was the pope's free port," he said, by way of explanation for the building's extraordinary size and undoubted expense. The Ancona lazaretto was designed to quarantine up to two thousand people at a time, in an attempt to wrest lucrative trade from its Adriatic competitors. Its floor plan follows the plague-containment template established at Dubrovnik and Split: arcades and courtyards for airing goods on the ground floor, suites of rooms to keep different groups of travelers separate above, and limited access points surveilled by guard towers to prevent escape. But, by the time construction was complete in 1743, Pugnaloni told us, "the plague was finished, so they never used it for that."

When Pugnaloni first visited, it was being used as a

tobacco warehouse. Now mostly restored, the unusual building hosts a bar (called the Lazzabaretto), galleries, event spaces, and offices. "I come to the theater here often," Pugnaloni said. Structures of quarantine, we were beginning to realize, are almost always built to contain the previous epidemic, often resulting in lazarettos that are obsolete before they open.

At the center of the vast pentagonal lazaretto in Ancona, Italy, is a structure intended for use as a pulpit from which to conduct religious services for those in quarantine. Beneath it is a freshwater cistern. (*Photograph by Geoff Manaugh*)

One of the most striking details Pugnaloni showed us was a large outdoor temple structure at the very center of the pentagonal lazaretto: it resembles a neoclassical folly on the grounds of an English stately home, but, in fact, performed two essential functions. Directly below the temple was the building's freshwater cistern, which was fed by an underground network of pipes connected to grates all over the surrounding courtyard, contributing to the facility's sanitary self-containment. But the structure had also been designed as a speaking platform for priests: an ingenious solution to the problem of conducting Mass for hundreds of potentially infected individuals without coming into physical contact with any of them.

In lazarettos from Florence to Omaha, the conflicting imperatives of religious congregation and quarantine have inspired acoustic, spatial, and technological innovations such as Ancona's courtyard temple. Indeed, in Split, Snježana Perojević had shown us a small pulpit mounted on the exterior of the southwestern tower of the lazaretto complex, from which a priest could perform an open-air, distanced mass for sailors quarantined aboard their ships. Providing spiritual comfort by whatever means possible to those undergoing the forty-day purification process of quarantine was, for much of history, perceived as having the same importance as ensuring their physical health and well-being. (The tradition continued in April 2020, when a priest in a town near Detroit achieved brief internet fame after he used a plastic squirt gun to anoint his parishioners' Easter baskets with holy water from afar during the COVID-19 pandemic.)

Our time in Ancona limited, we said goodbye to Pugnaloni, hurried back to the port, and continued up the coast toward Venice. There, we were scheduled to meet with a former high school history teacher named Gerolamo Fazzini, who had almost single-handedly fought and won a thirty-year battle to save the city's Lazzaretto Nuovo, or "new lazaretto," from developers.

Despite its name, the Lazzaretto Nuovo, which began construction in 1468, is one of the oldest remaining quarantine stations in the world. (The city's Lazzaretto Vecchio preceded it by half a century.) It occupies an entire island in Venice's eastern marshes, commanding the waterway into the lagoon. Fazzini's interest in preserving the island and its structures, we learned, originally came from a desire to re-use the facility as a martial arts training facility, but he soon abandoned that idea. Historical preservation in and of itself, Fazzini told us, was "more important than judo." Despite the island's immense historical significance, the lazaretto was, at the time of our visit, not yet open for tours; although it is listed as a destination on one of the city's regular ferry routes, we learned that our boat would not stop there unless we specifically requested it. A crew member watched us with amused curiosity as we stepped off onto the lazaretto dock and the ferry pulled away, leaving us stranded there, staring at a locked gate. Within a few minutes, however, Fazzini and an assistant emerged from a small gatehouse, unlocked the barrier, and welcomed us onto dry land.

Howard's own arrival at the Lazzaretto Nuovo was less cordial. "I was placed, with my baggage, in a boat fastened by a cord ten feet long to another boat in which were six rowers," Howard writes. Maintaining a distance was important: for all Howard or his Venetian hosts knew, he was infected. "When I came near the landing-place, the cord was loosed, and my boat was pushed with a pole to the shore." Now run aground, oarless, on the marshy shore of a remote island, Howard was fully committed to his fate.

Greeted by the island's prior, or head administrator, Howard was escorted across a large open meadow that we, too, now walked across, trailing Fazzini. Unlike us, however, Howard was led directly to what he describes as "a very dirty room, full of vermin, and without table, chair, or bed." The lazaretto, even in Howard's time, was ancient, having been built more than three centuries before he arrived. Howard tried repeatedly to wash the walls and floor of his tiny room, but reported little success. "This did not remove the offensiveness of it, or prevent that constant head-ach [*sic*] which I had been used to feel in visiting older lazarettos," he wrote. The prospect of spending forty days in such circumstances was a monstrous one.

Fazzini led us into the Tezon Grande, a huge warehouse used not for lodging but for storing and disinfecting goods. Thirty archways, now bricked in, stretched along its walls, extending twice the length of an Olympic swimming pool and making it the second-largest public building in Venice. Inside, fading graffiti decorated the whitewashed brick walls: a palimpsest of trademarks, coats of arms, Crusader monograms, drawings of ships, news updates, and travel stories. These marks had only recently been recovered from under the countless layers of lime applied to sanitize the structure between outbreaks. The graffiti, Fazzini explained, had been made using a syrupy dye of olive oil, brick dust, and rust, giving it a deep maroon color that resembled dried blood. One large inscription told the story of a ship's arrival from Cyprus in 1569; the looping, jumbled letters overflowed a wavy line of *denti di lupo*, or "wolf's teeth," drawn around the text as if to contain it. Another announced the death of the eighty-seventh doge of Venice and the election of his replacement—a gesture toward keeping the quarantined connected to the world from which they were temporarily separated.

With only pipesmoking, prayer, dice, or card games to pass the time, the urge to doodle, write, or otherwise make some kind of mark was evidently irresistible. From Venice to Sydney, those passing through quarantine have often left a more permanent inscription behind. "Over a hundred poems are on the walls / Looking at them, they are all pining at the delayed progress / What can one sad person say to another?" as "Xu, From Xiangshan," carved into the wooden walls of the barracks of the Angel Island Immigration Station in San Francisco.

Two rows of columns running down the center of the Tezon Grande were similarly decorated, but these marks functioned as a kind of inventory system, labeling specific bays, rooms, and spaces that held a particular company's goods. Ideally, the people and their goods would both cycle through forty days of quarantine at the same time so that crew and cargo could be released on the same day. In the 1500s, Fazzini told us, this hall would have been constantly perfumed with the scent of burning rosemary and juniper, used to fumigate the merchandise.

The lawns of the Lazzaretto Nuovo, or "new lazaretto," are lined with fragments of old pillars and the occasional olive tree. The bricked-up arches of the Tezon Grande are visible in the background. (*Photograph by Nicola Twilley*)

Outside, we walked across a lush green lawn, over to the twelve-foot-tall brick wall that surrounded the island, cutting it off from the marshy lagoon beyond. Although the priors who operated the lazaretto felt that green space was an essential amenity, Fazzini told us that, ultimately, most of the island was paved so that wool and animal skins could be aired out. "In the end, the disinfection of goods was felt to be of greater benefit to health than the tranquil spaces of the garden," concludes Jane Stevens Crawshaw. John Howard, with his characteristic attention to detail, listed the Venetian regulations pertaining to each kind of cargo, from beeswax and sponges ("purged by putting them in salt water") to ostrich feathers ("kept constantly exposed to the air, and very often moved and shaken") and tobacco ("ranged in heaps, and now and then moved"). These were, in his words, examples of "susceptible matter."

The Lazzaretto Nuovo had been built to expand upon and improve the quarantine capacity of the Lazzaretto Vecchio, explained Fazzini, but even all this extra space proved inadequate. In the plague year of 1576, which claimed a third of the city's population, hundreds of ships were anchored around the island to quarantine the overspill. To the Venetian notary Rocco Benedetti, the lazaretto looked like "Hell itself," with more than ten thousand people crammed on and around an island intended to house no more than a hundred or so, waiting amid the mosquitoes and humidity for the dreadful signs of disease to emerge.

Fazzini stayed on the island that night as we took a ferry back into Venice with his assistant, Ugo Del Corso. We learned that Del Corso was Italian but had grown up abroad. The Lazzaretto Nuovo—overgrown and dark, isolated and damp—had appealed to him the minute he saw it, he told us: it was the closest thing he had experienced here to his childhood in Indonesia. He said he would sometimes camp on the island for as long as two weeks at a time, alone, wandering around the weed-covered ruins, watching the sun come up, living for all intents and purposes beyond the edge of the world.

⟷

After just one night at the Lazzaretto Nuovo, 250 years before our own visit, John Howard caught a break. The next morning, he received a health inspection and was pronounced free of the plague. He was given permission to move across the city to the Lazzaretto Vecchio, or "old lazaretto," where he hoped he might experience a less extreme version of Venetian quarantine. "I hoped now to have had a comfortable lodging," Howard wrote. But his luck ran out.

Howard arrived at the Lazzaretto Vecchio—established in 1423, it was the first permanent plague hospital in the world—after a long boat ride. The lazarettos are separated by only fifty years but lie on opposite sides of the city. He was escorted into yet another room that repulsed him. "The walls of my chamber, not having been cleaned probably for half a century, were saturated with infection. I got them washed repeatedly with boiling water, to remove the offensive smell, but without any effect." He exerted the effort of contacting the British consul to arrange for his room to be whitewashed with lime, after which, he reported, "My room was immediately rendered so sweet and fresh, that I was able to drink tea in it in the afternoon, and to lie in it the following night."

The Lazzaretto Vecchio, or "old lazaretto," in Venice, Italy, is surprisingly close to its neighboring island, the Lido. (*Photograph by Nicola Twilley*)

The Lazzaretto Vecchio, we would see for ourselves, is an odd place: it is a natural island so thoroughly augmented by architectural construction that it now resembles a giant, partially submerged brick building. The island is walled off on every side, but, in its ruined condition, it is also open to the elements at every turn. We spent most of a day stepping in and out of collapsing buildings, past restored wooden boats filled with spiderwebs, the skyline and modern cruise ships of Venice never far from view. The island is separated from the Lido, a long barrier island at the mouth of the Venice lagoon, by a mere two hundred feet: it seemed so close to the rest of the city that it was hard to imagine why medical authorities ever tried to isolate people here. In part, we learned, this was because quarantine, in its earliest form, was much more of a collective experience than we currently imagine it to be.

In her book *Plague Hospitals*, Jane Stevens Crawshaw describes the city's sanitary efforts as an attempt at "regulating morality, behaviour and the environment in the fight against epidemic disease," with quarantine itself "an attempt to reintroduce order in times of plague." During an outbreak, the chaos created by widespread mass fatalities and what would very likely have appeared to be imminent social breakdown could be held at bay only by public health measures. These included regular religious services, the orderly passage of time marked by church bells, and visible, defined attempts at containment represented by quarantine.

As Crawshaw points out, the economic expense and logistical difficulty of "a decentralised system of household quarantine" quickly became too much for Venetian authorities. Concentrating and enclosing the potentially sick in one specially designed location was not just a rational medical solution; it was also a deliberate public relations effort. Just as Tomić and Blažina showed in Dubrovnik, quarantine was a sign that effective governance had been achieved. Lazarettos were "protective spaces," in Crawshaw's words, confining a group perceived to be both vulnerable and dangerous, offering both care and defense. Lazarettos made the city's fight against an otherwise insidious enemy visible.

The sick were separated from but remained closely tied to the city. As Crawshaw writes, "Families, districts and parishes were often quarantined at the same time." During a large outbreak, the cost of quarantine was borne by the entire municipality. Food, water, accommodation, and medical care for thousands of needy individuals were provided "at the expense of the state in the name of the godly Republic and the public good." The lazarettos became one of the so-called pious institutions of the city, to which, Crawshaw explains, it was compulsory for notaries to ask their clients if they would like to make a bequest in their wills.

Between visitations of disease, however, the lazarettos served as a barrier, shielding the city from external threats. Merchants and visitors covered the not-inconsiderable costs of their stay in quarantine themselves. By the time of John Howard's visit to its lazarettos, Venice had not experienced a significant outbreak of plague in 150 years, and the republic's quarantine ethos had shifted decisively, from a civic to a commercial endeavor. The value of a communal quarantine experience had also been superseded by a growing awareness of the benefits of isolation, and the lazarettos—by then already more than three hundred years old—had been crudely retrofitted to provide private rooms and a greater degree of separation.

Howard was unimpressed. "The *Venetians*," Howard wrote, "were formerly one of the *first* commercial nations in Europe, and the regulations for performing quarantine in their lazarettos are *wise* and *good*; but now, in almost every department into which I had opportunity to look, there is such remissness and corruption in executing these regulations, as to render the quarantine almost useless, and little more than an establishment for providing for officers and infirm people."

This was nothing, however, compared with what Howard would experience in Malta, our next destination. We boarded a short-hop flight from Venice to Valletta, a European city farther east than Venice and Rome and farther south than the capital of Tunisia, on our way to see a quarantine station that had helped to define an outer medical edge for the British Empire.

⟷

Malta is one of the strangest—and most eerily beautiful—places in the world. Almost as close to Africa as it is to mainland Europe but nevertheless a member of the European Union, the Maltese archipelago has been discontinuously inhabited since at least 5900 BCE. For a period of nearly one thousand years, it is believed, Malta's earliest culture entirely collapsed and the islands were left empty; their repopulation only came around 3900 BCE in the form of seafarers from Sicily. Malta's bizarre megalithic architecture, combined with deep parallel grooves eroded into the landscape known as "cart ruts"—which some archaeologists have interpreted as indicating the prehistoric use of heavy wheeled vehicles—have convinced thousands of armchair historians that Malta is the real-life site of Plato's Atlantis, an ancient civilization that is supposed to have sunk beneath the waves in an unidentified catastrophe. Even Malta's language is unusual: it is a Latinized version of medieval Arabic.

Like Dubrovnik, Malta is home to a large and distinctive quarantine facility. The Maltese lazaretto began construction in 1643, on Manoel Island, in the harbor just across from Valletta. It has hosted everyone from the Romantic poet Lord Byron to members of the British Parliament, from the Crown Prince of Abyssinia to the Italian painter Caravaggio. Byron, in a fit of boredom—or perhaps just arrogance—wrote his own name on the lazaretto wall. His graffiti was later encased in glass to preserve it, where it supposedly remains today, even as the wall on which he scribbled collapses on either side. (In a May 1811 poem called "Farewell to Malta," Byron waxed lyrical about his time in medical detention: "Adieu, thou damned'est quarantine, / That gave me fever, and the spleen!")

Our access to the otherwise off-limits site was organized by a local architect and heritage-consultant named Edward Said. In addition to his role as the curator of a vast eighteenth-century star fort immediately beside the lazaretto, called Fort Manoel, Said had already been working on the restoration of Manoel Island for several years when we met him. As he prepared to unlock the first of three gates for us, he mentioned one of his own pet theories, one that managed to tie together our recent travels. The architect of the Malta lazaretto, Said speculated, must have been to Dubrovnik. The organization of the interior rooms was too similar to be coincidental, he continued, and the staircases in Malta seemed inspired by the lazaretto outside the Ploče gate. A shared set of design principles for quarantine had become established, iterated across the seas and the centuries, despite the lack of any formal guidelines.

We continued through the first gate, asking Said about the status of the Manoel Island redevelopment plans. Said began an answer that involved the British architect Lord

Norman Foster—then he paused. We had reached a second fence, but something was wrong. Whether he had the wrong keys or the padlock had rusted beyond repair, we could proceed no farther. There were still two more chain-link fences between us and the lazaretto. We were locked out.

The sun was going down as we faced the possibility that we had come all the way to Malta only to miss our lone chance to get inside the island's legendary quarantine hospital. In the back of our minds was John Howard, who, when the French refused him entry to their lazaretto at Marseille on the grounds that its highly respected quarantine procedures constituted a trade secret, finagled his way in by posing as a doctor, thus securing a covert account of the building's inner workings.

We made an executive decision. Said insisted on staying behind, to ensure that security guards did not tow his car or lock us in, but the two of us quickly hopped onto the fence and climbed over. We came to another fence and climbed it,

A path through the vegetation slowly consuming an old staircase inside the lazaretto on Manoel Island reveals the passage of urban explorers. (*Photograph by Geoff Manaugh*)

too. In short order, we found ourselves alone, the only human beings inside Manoel Island's once legendary lazaretto, our time there limited not only by the setting sun but by Said's patience.

The building, we knew, was no longer structurally sound. We had, in fact, been warned against entering certain parts of the complex lest they collapse beneath us; later, as we jogged up ruined stairways and walked single file along crumbling balconies, we realized that no one had told us *which* parts. Invasive vines, erosion, and the effects of World War II bombing raids had clearly all taken their toll. Honey-yellow limestone blocks still stood atop one another, forming massive and imposing walls, but they no longer locked flush at their seams. Everything was slipping. In corners and crevices, weeds had begun to grow. At one point, we turned the corner into an interior courtyard to find ourselves face-to-face with a grand staircase, a diagonal path cleared through the overgrown vegetation carpeting its steps by quarantine tourists—or urban explorers—before us.

Maltese history is intimately tied to hospitalization, long before the age of quarantine. The Order of St. John, a military religious order and medieval fraternity of Christian knights who ruled Malta for centuries, are also known as the Knights Hospitaller. The Hospitallers played an early, Red Cross–like role as healers in the Holy Lands, a calling that began during the First Crusade, in eleventh-century Jerusalem. The failure of the Crusades—or, from a different perspective, the success of Muslim defenders in repelling barbaric European hordes—pushed the Hospitallers first to Rhodes, then to Malta.

In retrospect, Malta's generations-long transformation into an epicenter of European quarantine seems preordained. The Knights Hospitaller had, for centuries, been positioning themselves as warrior-physicians attending those on journeys far from home. In the 1800s, steamships transformed trade and travel, transporting infectious diseases from the Middle East and North Africa to Western Europe faster than ever before. The British were looking for a way to keep such exotic afflictions at bay without the fixed costs and maintenance associated with running a permanent quarantine station. Manoel Island sat at the center of that Venn diagram: by the 1830s, it had become Europe's busiest lazaretto.

For centuries, lazarettos had housed local populations during outbreaks of plague, and merchants, soldiers, and sailors the rest of the time, but, by the 1780s, when John Howard arrived in Malta, the demographics of quarantine were changing. The historian Alex Chase-Levenson quotes the Maltese quarantine superintendent Emanuele Bonavia pleading, in 1837, for funds to repair and upgrade the lazaretto, "as more and more people were arriving to perform quarantine, 'many of them ladies and generally all persons of condition.'"

These travelers paid for their stay—Chase-Levenson notes an endless series of fees for harborage, doctors, food, furniture, fumigation, and more—and, increasingly, their quarantine experience was calibrated according to their budget. An 1842 painting shows two Hungarian aristocrats in their private rooms in the lazaretto at Malta: they are lounging next to a fireplace in armchairs, potted plants on the mantelpiece, with room to spare. Crew members, who couldn't possibly afford to quarantine in such style, remained on their ships. "The crowded shipboard conditions from this period are well known, and the infuriating experience of remaining in them at voyage's end (and in sight of the shore) can only be surmised," writes Chase-Levenson in his book *The Yellow Flag: Quarantine and the British Mediterranean World, 1780–1860.*

For the British, who formed an increasingly large percentage of the lazaretto's population, Manoel Island offered a way to impose quarantine elsewhere. Britain's lack of a national quarantine facility was one of John Howard's most prominent complaints in his report from 1789, published under the title *An Account of the Principal Lazarettos in Europe*. "One consequence of my inquiries," he wrote, "has been a full conviction of the importance, to this country, of properly constituted lazarettos; and this, too, for commercial reasons, of which, I confess, I had before no idea." Much to Howard's dismay, British authorities insisted on treating quarantine as a burden that other countries should bear, in effect outsourcing the U.K.'s health border. Malta, situated between Sicily and North Africa and serving as something of a dividing line between the eastern and western halves of the Mediterranean, could not have been better placed: it is an outpost, an edge, and a place of encounter all in one.

Earlier that day, before our sunset scramble around the ruins, we spoke with Nicolina Farrugia. Farrugia, who spent her career in nursing, was only twenty-two when she started work on Manoel Island in 1970, in the final years of the lazaretto's operation, long after its heyday as a British outpost. As we talked, she told story after story of her experiences there, from running over the building's already-collapsing roofs to get from one ward to another, to working totally alone on the night of December 31, 1970, when a planeload of Libyan travelers landed and needed to be put into quarantine. "I could write books about this," she said, laughing, at one point. "Even now, I am getting gooseflesh!"

As late as the 1970s, the island had an air of isolation and remoteness despite its location in the heart of Valletta's harbor, Farrugia said. Her friends told her she should be afraid to walk to work on her own, across the bridge to Manoel Island then along the seafront to the isolation hospital. "But I was never afraid," she said. "I loved my work."

Both Farrugia and Dr. Herbert Lenicker, the last medical

director at the lazaretto, described their time there as one of gradual decline, with patient numbers dwindling as quarantine fell out of favor. Closing the lazaretto, Lenicker told us, was a mistake. "Number one, we still do get infectious diseases that need to be properly isolated," he said. "And, number two, I'm sure there is still a need for a quarantine—somewhere to put people and to observe what happens. There's a lot of risk in the way people move about nowadays."

As we explored the ruined lazaretto, we pictured Farrugia working night shifts alone in this vast complex of buildings, stairways, and courtyards—that, even in her time, were in a state of decay—and wondered whether John Howard had perhaps walked down these particular hallways, or stood on this exact terrace. We knew that, when Howard was in Malta, what he saw at the lazaretto did not inspire confidence. The "gloomy" internal spaces of the quarantine hospital, he wrote, were "so dirty and offensive as to create

On Malta, we had to climb two fences when our guide's keys did not work. Our reward was the opportunity to explore the old lazaretto on Manoel Island alone at sunset.
(Photograph by Geoff Manaugh)

the necessity of perfuming them . . . The use of perfume I always reckon a proof of inattention to cleanliness, and airiness."

Worse, the lazaretto staff seemed to have no interest in the well-being of their charges. People detained there, Howard wrote, "were served by the most dirty, ragged, unfeeling and *inhuman* persons I ever saw. I once found eight or nine of them highly entertained with a delirious *dying* patient." In the end, the lessons of quarantine cannot be reduced to architecture; Howard's travels revealed that even the most well-designed facilities, purpose-built to handle both potentially infected merchandise and prolonged human medical confinement, can be infernal if not run with care.

Howard tried to summarize and express everything he had learned about quarantine in his own speculative design for an ideal lazaretto. Located on a hypothetical spit of land overlooking the sea, and situated amid cleansing crosswinds, Howard's proposed structure would have been divided into wards dotted with independent buildings and gardens. All of these, in turn, would be arranged around a spacious central green. "A lazaretto should have the most cheerful aspect," he wrote. "A spacious and pleasant garden in particular, would be convenient as well as salutary." Howard included a chaplain's house, a home for the lazaretto's inspector, and sheds for purifying goods, among many other specifically appointed spaces. He placed a burial ground outside the lazaretto wall near an "area for foul goods." If someone stood there, looking down at the harbor from between the graves and infected cargo, they would have seen a separate "landing place for foul ships" opposite a "mooring for safe ships," and a long iron chain strung from one side of the harbor to the other, to prevent stray arrivals—or escapees.

This vision, though never built, gave topographical form to Howard's experiences with quarantine, offering future

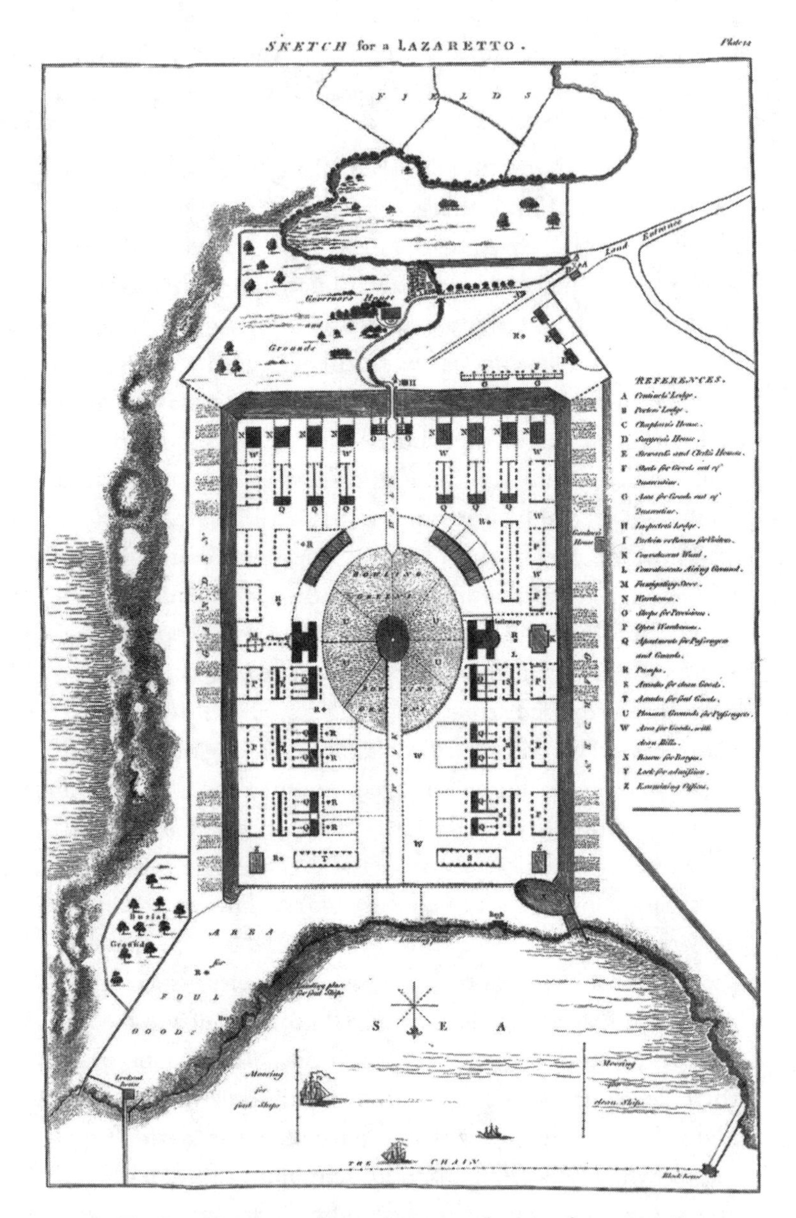

The British prison-reform advocate and critic of quarantine John Howard drew up plans for his own "ideal lazaretto." The facility was never built. (*Courtesy of the Wellcome Collection*)

detainees and suspect travelers the cleanliness, personal space, and communal leisure areas that he had longed for in his journey through European medical detention. With Howard's extreme attention to detail, the plan also sought to ensure protection from disease through orderly and well-planned circulation.

Howard's ideas for prison reform were carried forward by others and ultimately created practical improvements in jail design and operation. His thoughts about quarantine had less of an impact. His English readers, rather than reacting with horror at the unexpectedly dreadful conditions of British prisons, were more likely to find that Howard's report confirmed their suspicion of Continental practices. British politicians and merchants remained convinced that whatever inconvenience accompanied the country's reliance on Malta, or on disused ships moored offshore, for its quarantine capacity was more than outweighed by the resulting cost-savings and flexibility.

Despite the risks, expense, and discomfort of his travels, and the thoroughness of his voluminous report and statements to Parliament, Howard did not, in the end, achieve his goal: the construction of a British national lazaretto. It is perhaps indicative of quarantine's general lack of appeal to British authorities that even the facility approved for construction on Chetney Hill, a marshy island in the Thames estuary, was abandoned, its materials sold at a loss, its remnant walls left to sink into the marshes. "Since then," a historical report published in 1964 explains, "the island has reverted to pasturage; foundation lines with remnants of ruined walls . . . are all that now remain of this unique and costly episode in British preventive medicine."

Nonetheless, in his observations, Howard helped to identify essential elements of a well-run quarantine facility, from ample ventilation and carefully designed circulation to communal religious services and compassionate care—details that planners, engineers, and health officials are, sadly, still getting wrong today. As we discovered, his itinerary has much to offer a twenty-first-century quarantine tourist hoping to understand the origins and development of this powerful, often-misunderstood public health tool.

⟷

In July 1789, Howard embarked on his final journey, hoping to travel by land deep into Eastern Europe, entering Russia. He was motivated by a desire to learn more about the origins of the plague, and, according to his friend and frequent coauthor Dr. John Aikin, to "collect some information respecting it that might lead to a discovery that would arrest its progress." Howard sent his last letter home in September 1789, having found himself in the midst of military conflict between the Russian and Ottoman armies. In the letter, he wrote, "My spirits do not fail me; and, indeed, I do not look back, but would readily endure any hardships, and encounter any dangers, to be an honor to my Christian profession."

While Howard was staying in a Ukrainian town called Kherson on the shores of the Dnieper River, a group of Russian soldiers approached him. They had learned of Howard's long experience inspecting hospitals and lazarettos, and had come to him for medical advice. A young woman in a town twenty-four miles away had come down with an illness of some sort after attending local Christmas festivities. Although not a medical doctor, Howard agreed to help; he rode off on horseback, accompanied by an admiral of the Russian army.

The woman, Howard saw, was suffering from typhus—the dreaded gaol fever. Although it is impossible to know for sure, Howard believed it was this exposure, out of the thousands to which he had subjected himself, that finally penetrated his defenses. He soon came down with symptoms, writing in his diary that he suspected he had contracted the woman's illness while assisting her. Bedridden for nearly a month, feverish, in pain, and with a dry hacking cough, Howard died of the contagion on January 20, 1790.

While he is commemorated elsewhere solely as a prison reformer, in Kherson, John Howard is firmly associated with quarantine. A story that still circulates in Kherson today claims that Howard recommended any residents suspected of suffering from typhus to be isolated on what became known as Quarantine Island, in the Dnieper River. Howard's advice is credited with saving hundreds of lives and ensuring the town's survival. A 1945 article in the *Soviet News* relates that "generation after generation were told about the Englishman who came from far off to help them in time of their distress."

3

Postmarks from the Edge

As a young boy, evacuated from London during the Blitz, Denis Vandervelde realized he faced two major obstacles to achieving distinction in the field of philately: he was penniless, and, worse, color-blind. "A lot of the expertise in stamp collecting depends on shades and tiny variations in printings," he told us. "So it was a daft hobby for me to have."

Instead, Vandervelde began collecting postmarks and quickly found himself absorbed in an entirely different pursuit—a kind of postal treasure hunt, documenting the elaborate bureaucracy that has emerged to manage the movement of mail around the world. Stamp collectors prize misprints and rarities: the most valuable stamps in the world include a green stamp that was meant to be pink, a sheet accidentally printed without perforations, and the "Inverted Jenny," a twenty-four-cent stamp featuring an upside-down biplane. Similarly, postal historians particularly relish the markings generated in response to exceptional circumstances: censored mail, siege mail, and "wreck mail," retrieved from train or plane crashes.

Vandervelde's own collection, assembled over nearly fifty years, now boasts three thousand items of disinfected mail: suspect letters that have been punctured, perfumed, or otherwise purified to prevent the transmission of disease.

Together, they trace the outline of a parallel pandemic infrastructure, one whose contours have, in many cases, hardened into the passports, borders, and institutions that still control global movement today. By studying the traces left by disinfection—scorch marks, stains, and incisions—and the distinctive cancellations used to mark mail as treated, Vandervelde and his colleagues in the Disinfected Mail Study Circle, an international group of hobbyists and collectors, have ended up performing a forensic archaeology of quarantine through its postal paper trail. On the one hand, these letters provide the evidence that allows them to reconstruct outbreaks of both disease and its traveling companion, fear. On the other, mail acts as a proxy for cross-border flows of people and goods. Its treatment establishes documentary evidence of both permanent and pop-up systems of infection control.

We met in Islington, London, at Stampex, a biannual philatelic fair that is Europe's largest. Vandervelde, a sprightly octogenarian, suggested that we join him after a morning's booth-browsing as he refueled with a glass of Shiraz and a pizza at the convention center café. Over lunch, he explained that, as with human quarantine, the practice of disinfecting mail was first formalized in the Adriatic, though no one knows exactly when. Certainly by the 1490s, following a century during which Venice had been scourged by a fresh outbreak of plague every decade, that city's health authorities had decided that it might be wise to extend their sanitary precautions to letters that came from infected or suspected areas, not just people. As with many of the other practices of quarantine, including the institution of the lazaretto itself, other port cities were quick to follow in Venice's footsteps.

"You have to remember that the general belief at the time was that all infectious diseases were a miasma—a kind of a cloud that could attach itself to things," explained

Vandervelde. "Therefore anything could be subject to infection." Items that can harbor and transmit contagion are called *fomites*, from the Latin for tinder—they can kindle disease. Not everything was considered equally conducive to conveying miasma, however: soft materials, such as cloth, wool, and even fruits and vegetables, were thought to be highly susceptible, whereas hard objects, such as wood, metal, and tortoiseshell, were seen as impervious to infection.

Paper sat in between these two extremes—it was theoretically susceptible, but not especially likely to carry disease. The degree of risk represented by a letter was a matter of the finest discrimination, Vandervelde told us. "For example, in Venice, if the mail was bound with a linen thread, which it quite often was, it was definitely dangerous," he said. "But if it was bound with wire, which increasingly they did, it would quite often be allowed through without treatment."

Before the eighteenth century, official medical records are sparse: disinfected mail inadvertently provides a time-stamped, geotagged archive of contemporary epidemic intelligence. Even when the precise date or location of treatment is missing, Vandervelde can often deduce it from, for example, regional variations in fumigation technique or the adoption of new disinfection technologies.

"In the early days, the mail was simply put into a wooden coffin with sweet-smelling herbs and spices," explained Vandervelde. "It had to be in there for at least a week, and, if it wasn't collected in six weeks, it would be destroyed." Later on, health commissioners in the Mediterranean lazarettos adopted a process they called the *spurgo*, or "purge"—"a much more violent treatment with vinegar and smoke." In all cases, the logic was that these strong odors were capable of displacing any disease-ridden bad air that might have impregnated the paper en route. Letters were sprinkled or dipped in vinegar, leaving distinctive splash marks; they were then put on a wire grate and grilled, or held over a fire with tongs, whose ghostly impressions remain visible as white lines on a browned envelope. Health officials in some locations were known for their particularly enthusiastic disinfection, with correspondence that passed through Marseille often emerging unreadable.

Brittle, stained, discolored, and often adorned with an official cachet or wax seal, the letters reassured recipients of their mail's safety—on the outside, at least. Indeed, many Italian health authorities drew attention to the limitations of their disinfection, with a stamp that read *netta fuori e sporca dentro*, or "clean outside, dirty inside." "You may well ask what on earth you were supposed to do when you received a letter like that!" Vandervelde said with a laugh. "Open it?" He told the story of a much later outbreak, when a circus entertainer from India brought smallpox to Launceston, Tasmania, in 1904. In response, the Australian postal service disinfected the city's mail for three months, marking thousands of letters as treated. "Ninety-nine percent of recipients put those letters straight in the fire," said Vandervelde. "Those marks are really quite rare—there are about fifteen of them known."

Later, when we repaired to his semidetached house in the north London neighborhood of Golders Green to spend an afternoon marveling at the highlights of his collection, Vandervelde showed us one of the earliest known examples of internal disinfection. "It's from Naples, where they were very fierce," he said. "As far as I know, they were the first to slit." The letter itself is from a missionary stationed in Bulgaria, reporting back to his superiors in Rome. Based on the contents, which Vandervelde had translated, the man's mission had not been a success. "Few know the Paternoster or the Ave Maria," the missionary wrote, in looping cursive. "And they refuse to be taught. I have had to bribe the children. Pray for my salvation."

As charming as the missive's contents are, Vandervelde's real interest is the exterior, which has been slit open with a chisel at each corner to allow the fumigant inside. Although the letter itself isn't dated, the Neapolitan health authorities replaced their handstamp so frequently that Vandervelde can narrow it down to between July 1755 and September 1756, based on which one of half a dozen variants it is. "I'll have you know I picked this up as junk," said Vandervelde, with pride. "It really is a lovely, lovely letter."

A "plague apparatus" from one of the Venetian lazarettos, used for disinfecting mail and other paper goods. (*Courtesy of the Wellcome Collection*)

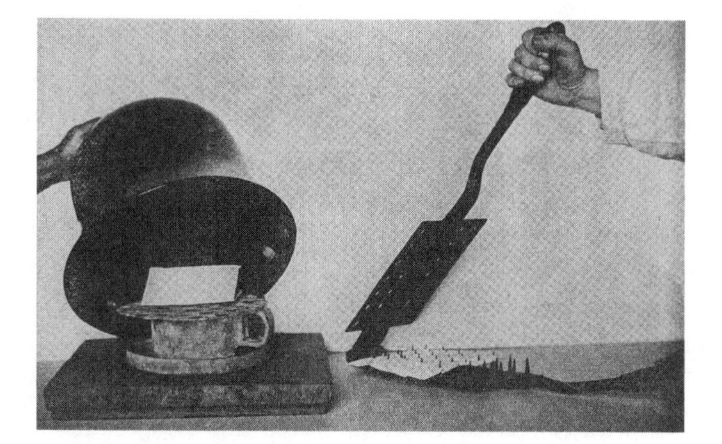

Mail disinfection equipment; the device being demonstrated on the right is a rastel. (*Courtesy of the Wellcome Collection*)

By 1787, France had adopted this new internal disinfection technique—Vandervelde has a letter from that year, with "quite serious" chisel marks—after a chemist wrote to Louis XVI, who was soon to be guillotined himself, to let him know that the kingdom of Naples was a step ahead. Later, the practice of slashing mail with chisels and awls, which often left it in shreds, was made obsolete by a device called a *rastel* (from the Latin *rastellus*, or "rake"), which resembles the love child of a waffle iron and a medieval torture device. Letters were placed between hinged, spiked plates and punctured pre-fumigation—the particular pattern of holes punched through the paper, from the offset grid used in Hamburg, Germany, to the distinctive sunburst of Mahón, Minorca, can provide disinfected-mail collectors with yet another clue. "In Trieste, I've got proof that at least four different machines were used," said Vandervelde. "Always with one row of slits that way and one row of slits the other way"—but each with its own distinctive imprint, or tell.

The study of disinfected mail began in the 1950s. In Trieste, an Italian urologist named Carlo Ravasini began collecting disinfected letters from Italian city-states and publishing on the topic. At about the same time, in San Francisco, a Swiss-American scientist named Karl F. Meyer saved the California canning industry with his research on botulism before making himself a World Health Organization authority on plague. Meyer's work included studying disease transmission in animals by constructing a miniature simulated "Mouse Town" surrounded by a white-painted moat filled with crystals of DDT to prevent the escape into San Francisco of any Black Death–carrying fleas. One day, "in a period of temporary ennui," as he later wrote, Meyer came across an 1898 letter stamped "MIT FORMALIN DESINFICIERT" ("disinfected with formalin"). He was intrigued. As he later wrote, an "insidious sideline grew into a major addiction": rarely a month passed "without the exhilarating pleasure of finding a rare new item in an auction," resulting in bulging albums and, in 1962, the publication of *Disinfected Mail,* still the only comprehensive book on the subject.

When a young Denis Vandervelde acquired his first disinfected letters, almost by accident—the auctioneer had put his daughter in charge of calling the bids, and she teased the shy youth into blowing his entire budget of £6 on two Italian examples with interesting markings—he began to research the topic by reading Meyer's book. He then wrote to Meyer, expressing his interest in the subject. "He wrote back a letter of fifty-nine pages," said Vandervelde. "And I was wading my way through this when I got a telegram from him saying, 'At age eighty-eight, I cannot afford to wait for answers.'"

Vandervelde quickly apologized, and the next time Meyer was in London, on his way to Geneva for a meeting of the World Health Organization, the two of them had dinner. They became fast friends, though Meyer died just before Vandervelde founded the Disinfected Mail Study Circle in 1973. "I've got a letter from him saying you'll be lucky if you get more than six members," said Vandervelde. "We're now a hundred and fifty strong, in twenty-five countries, and still growing." He estimates that at least a third of DMSC members are medical historians or doctors, a third are collectors, and another third are stamp dealers or authors who write about related subjects. (We joined the Circle in 2009, swelling the ranks of authors by two.)

Until recently, when Vandervelde "retired" for the second time, the benefits of membership included a subscription to *Pratique*, the society's newsletter. It is named after the permission, called *pratique*, given to ships to dock and do business after demonstrating their freedom from contagion, either through the performance of quarantine or by presenting a valid bill of health. Each issue, although occasionally many months delayed due to technical difficulties that often involve Vandervelde's antiquated AOL account, is filled with stories and scenes that convey the flawed reality of quarantine, despite the ambitious health regulations governing it.

A letter bearing a simple blue handstamp from Boston's quarantine station on Rainsford Island, for example, is accompanied by an ominous excerpt from the *Boston Almanac*: "Feb. 5, 1836—The inner harbor frozen over, and a pilot walked from Quarantine Island to the city on the ice." Vinegar-doused envelopes from the Ionian Islands—"always very suspect," noted Vandervelde—serve as the backdrop to a story in which a plague outbreak on Cephalonia is traced back to the local Catholic priest's recently deceased illicit son, who stole flea-infested clothing from the corpse of a Turkish army officer while harvesting wheat overseas as a migrant laborer. The climax of the tale involves the British officer in charge of enforcing quarantine ordering the priest to dig up his earthen floors to reveal the dead son, hastily buried in his magnificent but deadly new clothes.

If the letters themselves are pieces in a puzzle, Vandervelde and his colleagues are obsessive in their desire to fill in the gaps, collecting maps and tracking down contemporary sanitary proclamations that shed light on the slightest variation in quarantine practices. Over coffee and cookies in his book-lined, thickly carpeted living room, Vandervelde told us about a series of ship's papers he acquired thirty years ago, when a DMSC member told him about a cave in Switzerland whose entrance had recently been rediscovered. The naturally temperature-controlled cave was stuffed with old, perfectly preserved documents, and the family who owned the land was selling boxes of this paperwork off in bulk. "Two-thirds of it was complete rubbish," said Vandervelde. "People's laundry lists, basically." But, astonishingly, amid the trash were a series of maritime bills of health that helped explain the routes taken by dozens of disinfected letters in his collection. Similarly, some of Vandervelde's most prized items come from an antique-furniture dealer in the English town of Swindon who discovered a huge collection of disinfected mail stuffed inside some of the pieces he'd obtained while cleaning out strangers' attics.

In painstaking detail, for missive after missive, Vandervelde has traced how, where, and why quarantine operated in the premodern world—a largely forgotten geography of lazarettos, rastel stations, merchant ships, colonial checkpoints, health passports, and border crossings. At times, his research has revealed outbreaks that had otherwise been lost to history: an 1897 outbreak of plague in India that triggered disinfection measures in southern Russia, and, for one day only, in New York City. (There, a paranoid postmaster decided, seemingly on his own initiative, to fumigate a bag of mail from India that arrived on the steamship *Britannic*. He even had a circular cachet made up to mark each letter: NEW YORK DEPARTMENT OF HEALTH, DISINFECTED, JAN 30, 1897.)

More often, the mail actually holds up a mirror to the epidemic of fear that so often accompanies infectious disease. Vandervelde showed us a cache of fifteen letters from the 1660s, all carried by the private Thurn and Taxis postal service, beloved by fans of Thomas Pynchon's novel *The Crying of Lot 49*. Taken together, these letters graph the rise and fall of prophylactic disinfection as news of London's Great Plague of 1665–66 rippled outward across Europe's cities.

Some of the letters in Vandervelde's collection were actually mailed from quarantine itself—even, in the case of one letter, dispatched from the heavy iron chain that had been slung across the harbor of Marseille to prevent ships slipping into port without pratique. (The letter's return address reads, simply, *La Chaîne*.)

One of Vandervelde's most prized items is the earliest known letter sent to or from quarantine in Britain. While continental European ports and border crossings fortified themselves with lazarettos, cordons sanitaires, and disinfection stations, England—to John Howard's disappointment—lagged behind in the adoption of quarantine measures. In the 1300s and 1400s, despite suffering from repeated and devastating outbreaks of the Black Death, the most severe of which killed an estimated 45 percent of the population, England was far from the seafaring, colonial empire built on trade that it would later become. Without a clear point of entry for disease, the country was reluctant to invest in the expansive—and expensive—public health infrastructure established in Venice, Ancona, Dubrovnik, and Marseille. Finally, in the autumn of 1710, rumors of plague in the Baltic ports led Jonathan Swift, best known for his later satire

Gulliver's Travels, to write to the chancellor of the exchequer, begging him "for the love of God to take some care about it, or we are all ruined." Parliament passed England's first quarantine act a few days later.

At first, English port authorities simply directed suspect ships to anchor in the Thames estuary, their goods aired out on deck, while passengers were made to wait out quarantine in sheds or in the run-aground hulks of disused warships. Vandervelde's letter is dated April 10, 1765, and mailed to the *Reward*, "lying quarantine in Stangate Creek." It's from one brother to another, clearly written in response to an earlier message in which the brother in quarantine had apologized for this additional delay on his journey back from İskenderun, Turkey, a coastal town near today's Syrian border.

But, while the letter provides documentary proof of the location of quarantine in England, Vandervelde had, until recently, been unable to tell where it was mailed from—the return address is simply given as Leigh, and there are fifteen different villages in England with that name. A few years ago, however, he gave a talk to local historians in the seaside town of Southend, during which he showed this letter. "An old lady in the audience got very excited," Vandervelde told us. "I thought she was having a fit."

As it turns out, the woman was writing a history of one of the Leighs—specifically, the ancient village of Leigh just inland from where the modern town of Southend was built as a resort—and she knew exactly who these brothers were. "The story was that the village of Leigh at this time was so poor they couldn't afford a vicar," Vandervelde said. Instead, a local farmer volunteered his four sons to rotate the job among them. Thanks to quarantine, the brother held at Stangate Creek was, in fact, late for his shift ministering to the spiritual needs of a few hundred poverty-stricken whelk, winkle, and cockle harvesters. (In 2018, Leigh was named the happiest place in Britain, so its fortunes have clearly changed.)

⟷

A jurisdiction establishes itself through its bureaucratic processes: the issuing of coinage and postage stamps is often the first order of business for a new state. (Indeed, many of the tiny, remote island nations in the Pacific—Niue, Kiribati, Tuvalu—depend on postage-stamp sales for a large proportion of their annual revenue.) Similarly, the otherwise imaginary and invisible line around a country—its border—was typically articulated through quarantine and other health-screening practices: disease management, by attempting to keep microbes *out*, defines the edges of *in*.

Thus, when the southern provinces of the Netherlands seceded, in 1830, to form their own country, Belgium, they immediately issued their own coins and stamps. They also introduced quarantine controls. Intriguingly, the borders of this new nation already bore the imprint of medical isolation: a few thousand feet south of the spot at which Belgium, the Netherlands, and Germany meet, a small, almost perfectly rectangular chunk of German territory, roughly the size of two American football fields, extends into what should logically be Belgium. The explanation for this territorial anomaly can be found in the grassy patch's name: Melatenwiese, or Leper Meadow. In the Middle Ages, the nearby city of Aachen banished lepers there, just outside the city walls. Although the land technically belonged to what would later become the Belgian province of Liège, a lingering fear of infection meant that no one exercised this claim, even long after the leper camp disappeared, leaving the empty field to become de facto—and then officially—German.

Among the sanitary measures introduced by the new Belgian nation was, of course, the disinfection of mail. "Wanting to be quite different from both the Dutch and the French, they decided that the way they would deal with mail was to open up each letter, smoke it, and then reseal it by sticking a label with an apology in French," Vandervelde said. This label—a three-by-two-inch rectangle of heavy cartridge paper, in a distinctive shade of gray—is Vandervelde's white whale. "I would dearly love to get a complete example," he said. Because the label was typically stuck over the fold, it was almost always ripped in half to get at the letter inside. "Consequently, there are only, as far as I know, three examples in the world of this thing complete, and two of them are almost certainly printers' samples," he said. The one that Vandervelde wants is stuck on the back of a letter that has never been opened: he saw it in an auction catalog after the sale had taken place, and has no idea where it ended up.

In addition to its rarity, one of the things that makes this Belgian disinfection label particularly significant is that 1830 also marks the first outbreak of cholera in Europe. The disease seems to have been endemic in the Ganges Delta region for centuries, but just four years earlier most Europeans had never heard of it. In an article for the *London Review of Books*, James Meek describes the Russian writer Alexander Pushkin playing chess with a friend who, Pushkin recounts, "knew a lot of the kinds of thing they study in universities while we were learning to dance," and who, while checkmating him, warned that "*Cholera morbus* is at our borders, and in five years, it'll be here."

That was in 1826. In September 1830, Pushkin left Moscow to visit his family's estate in the countryside for a couple of weeks, and ended up spending three months there, sheltering in place. Initially annoyed, he wrote to his fiancée, who had remained behind, to vent. "I've been notified that there are five quarantine zones set up from here to Moscow, and I will have to spend 14 days in each," he complained. "Do the maths and imagine what a foul mood I am in!" Nevertheless, he quickly settled into lockdown, growing a beard and mustache, living on potatoes and buckwheat porridge, and producing some of his best writing. Meanwhile, all around him, workers and peasants panicked, rioted, dowsed themselves in tar-water as a disinfectant, developed conspiracy theories involving poison and Poles, and were shot at by armed soldiers enforcing restrictive quarantine lines.

By 1831, cholera had reached Finland, Poland, and Austria. From there, it spread to the Baltic ports, arriving in Sunderland, England, by December 1831, in New York City and Philadelphia a year later, and in Mexico and Cuba soon after that. In Russia alone, more than a quarter of a million people died of cholera—or, more precisely, of extreme dehydration brought on by the profuse, watery diarrhea that is the disease's main symptom. Over the next sixty years, four more deadly cholera pandemics would sweep across Europe and the world, killing more people more quickly than any other epidemic disease of the nineteenth century. Meanwhile, European and American attempts to prevent its spread ended up defining the systems of global governance that still shape the world today.

Much had changed since quarantine was first formalized in the Adriatic. The disease in question was very different: cholera is spread by consuming food or water contaminated by a sick person's feces, rather than by fleas (bubonic plague) or person-to-person, by breathing (pneumonic plague). Still, until Robert Koch discovered the cholera bacillus in the 1880s, and Paul-Louis Simond showed, a decade later, that

Yersinia pestis is transmitted by fleas, many people still believed all three were spread by miasma, or noxious air.

More significantly, technological progress meant that, in the nineteenth century, the movement of people and goods underwent an exponential acceleration. The great lazarettos of the Mediterranean were built for an age of sail. In 1819, the first transatlantic crossing by steamship took place, and a journey that had taken four to six weeks under sail was reduced to nine days by midcentury. When the Suez Canal opened in 1869, it offered a shortcut that brought Asia and Europe nearly four thousand miles closer together. On land, railways crisscrossed continents: by the 1880s, the Orient Express whisked travelers from Paris to Constantinople in style—and in under three days.

Alison Bashford is an Australian historian whose work has examined the shifting lines drawn between clean and dirty bodies, spaces, and countries; she grew up in Sydney, but later moved, appropriately, across the harbor next to the quarantine station in Manly. She explained to us that Europe's fear of cholera became focused, in particular, on a suspected super-spreading event—the hajj. Since medieval times, hundreds of thousands of Muslims from the Middle East and Asia had undertaken this sacred pilgrimage to the holy city of Mecca every year, but their journeys—by camel caravan and by sail, subject to the vagaries of monsoon winds—took so long that any disease they might have been carrying would have revealed itself, and then burned out, en route.

By the mid-1800s, said Bashford, with railways connecting Persia and the Red Sea to the Mediterranean, "there was constant European anxiety that this mass movement—in a place that is just adjacent to Europe—would introduce cholera into Europe itself." Cholera, like plague before it, was seen as a disease that had been inflicted on the infidels of the East, and now threatened the Christian West. The boundary between the two—"which," Bashford points out, "was not an actual territorial border but, rather, an important part of how Europe defined itself in relation to its most adjacent neighbor, the Orient"—became the place where cholera was controlled, through a rigorous sanitary policing of Muslim movement. The metaphorical border between East and West was, Bashford told us, "made meaningful through very specific, very grounded practices of inspecting people, putting them in quarantine camps, and monitoring or restricting their movement."

One of these camps was located at El Tor, a tiny port at the tip of the Sinai Peninsula, where the Suez Canal opens onto the Red Sea. It was run by Egyptians under the direction of the European powers who had organized themselves into a sanitary and quarantine board with broad authority to ensure that ships arrived at Alexandria, on the Mediterranean end of the Suez Canal, free from disease. (The Egyptian authorities had initially requested public health advice from foreign consuls stationed in the country, not expecting or, indeed, welcoming the full international takeover of their quarantine system that resulted.)

According to Denis Vandervelde, who has a postcard mailed from this camp in his collection, conditions at El Tor "were primitive in the extreme." He showed us illustrations from *The Graphic* magazine that depicted a handful of wooden sheds built on the beach to house passengers, the distinctly spartan interior of the Ladies' Quarantine Shed, and the even more austere Cholera Tent, for those who fell sick. Of course, every time anyone did show symptoms, the duration of the quarantine was extended, meaning that pilgrims could easily spend months waiting on the beach, subjected to intense daytime heat and freezing cold nights, as well as a persistent north wind that blew sand into every crevice. (Conjunctivitis and other "suppurating" eye conditions were common.) Two cisterns were available for five hours a day, to serve three hundred or more people, leading one pilgrim to report "nights spent tormented by thirst." As the historian Patrick Zylberman puts it, pilgrims were perceived—and treated—as carrying a "double contagion": in addition to being disease vectors, they were Muslim, and thus part of a Pan-Islamic brotherhood considered threatening to European colonial rule.

This hajj-centered quarantine infrastructure marking the edges of Europe actually built on a longstanding land-based sanitary cordon: for a century, from 1770, Austro-Hungarian authorities maintained a thousand-mile quarantine corridor along their imperial frontier, all the way from the shores of the Adriatic to the Transylvanian mountains. This epidemiological boundary was not simply a line but a buffer zone, thirty miles wide in many places, cutting a broad swath through modern-day Serbia, Bosnia, and Croatia. Inside this belt, every peasant was also a soldier, responsible for manning the sanitary cordon for at least one week in every eight, and more during an outbreak—up to a total of six months' active service each year. A chain of two thousand lookout posts was constructed, each no more than a musket-shot's distance from the next, and soldiers were instructed to fire on any unauthorized traffic. Nineteen crossing posts offered disinfection services, open-air *parlatorios* for distanced conversation across the divide, and supervised quarantine for travelers—twenty-one days when no outbreak was suspected, and forty-eight when the presence of plague was confirmed in the region. The cordon served as both military and public health defense, protecting Europe from Oriental incursions of all sorts.

Although this quarantine corridor was dismantled in 1871, its geopolitical legacy continues to resonate. Europe's conceptual and institutional dissociation from its eastern neighbors in the former Ottoman Empire persists, as can be seen in the bloc's distinct lack of enthusiasm in response to Turkey's decades-long quest to join the European Union. Meanwhile, the Balkan buffer region, following a century of restricted mobility, forced immigration, and economic lock-down, is still a militarized and marginalized place whose unresolved but fiercely contested ethnic identities have fueled the region's recent bloody conflicts.

In a study published in 2019, a team of researchers found that inhabitants of the former cordon sanitaire territory are still poorer than their neighbors on either side, with measurably lower levels of interpersonal trust, and a distinct tendency toward bribery in their dealings with public institutions. As the journalist Jessica Wapner has written, the symptoms of *Mauerkrankenheit*, or "wall disease," include "a sense of being locked up and of being isolated from friends and family." The malady was first named by East Germans living in the shadow of the Berlin Wall, but, as researchers have subsequently discovered, the diagnosis is applicable to anyone living "in borderland regions," whose populations are typically characterized by higher levels of depression, suspicion, and poverty. "Walls immobilize us," writes Wapner. "This restriction can be devastating to our mental health."

More hauntingly, the borderlands were also ground zero for alleged vampire sightings, triggering a literary mania that swept Europe. This quarantine threshold, a zone of suspicion and uncertainty, whose inhabitants were neither healthy nor sick, neither citizen nor soldier, and constantly under threat from plagues both real and imagined, proved the perfect hunting ground for the similarly liminal living dead. In the most famous vampire story of all, Bram Stoker's *Dracula*, these monstrous creatures are "mutations at the periphery of the world—new forms, new creatures, new diseases—come back to haunt the world at its center," according to the literary historian Thomas Richards. The center, in this case, is London, where Count Dracula arrives by following much the same route as the Black Death: boarding a Russian vessel in the Black Sea and sailing through the Mediterranean, past Malta, to the U.K. Indeed, as if anticipating Denis Vandervelde and the DMSC, Richards point out that "because Dracula must hire intermediaries to carry him from place to place, his movements can be traced through invoices, memoranda, and other documents." Quarantine had infected the European imagination, inspiring its most potent horror stories, both fictional and real.

⟷

Back in Islington, as we waited for the check from lunch, Denis Vandervelde beamed, paper napkin still crumpled in his hand, as he posed for a photo with his most recent acquisition, a trophy from that morning's Stampex foray. The disinfected letter, sent in 1801 from Málaga, Spain, to what is now Ghent, Belgium, was the earliest postal evidence he'd found of the great European yellow fever scare.

Yellow fever, or "the exterminating disease of the Black Vomit," as Spanish conquistadores called it, was endemic in Africa, where the population had acquired some immunity. It was likely brought to the New World in ships transporting enslaved people. For Indigenous Americans, it was just one of a suite of new diseases imported by Europeans that devastated native populations, with a death toll that will never be known. Estimates of the pre-Columbian population of North America range from two to eighteen million; by the end of the 1800s, only half a million people were left.

For Europeans, yellow fever was a terrifying new affliction, "which took lives indiscriminately and mysteriously," according to Pedro Nogueira, a Cuban historian writing in 1955. "Certainly for centuries it appeared that the treasures of the tropics were guarded and protected by a monster." The British army lost twenty thousand out of the twenty-seven thousand soldiers sent to try to take Cartagena, Colombia, to yellow fever in 1741; roughly 10 percent of the population of Philadelphia was wiped out by the disease in 1793, leading to the construction of the Philadelphia Lazaretto—the first structure of its kind in the Americas. When Napoleon sent sixty thousand French soldiers to Haiti to quash a slave rebellion in 1801, 80 percent of them died within two years, jaundiced, feverish, and vomiting a noxious substance that resembled spent coffee grounds.

Defeated, Napoleon sold off the rest of his North American possessions to the United States in the Louisiana Purchase. Meanwhile, what was left of the imperial fleet sailed back to French and Italian ports in 1804, bringing yellow fever with them. But Vandervelde's envelope provides philatelic evidence of an even earlier outbreak, in the ports of southern Spain; the first sparks in a yellow fever panic that triggered disinfection and quarantine measures as far away as Moscow.

"Now, what the doctors should have spotted, and never did, was that no one living more than a half a mile from the seafront ever died in these ports," said Vandervelde—a distance that happens to be the flight range of a mosquito. "They didn't know it was a mosquito," he added. "But they really should have put two and two together."

It wasn't until almost a century later that an American military physician, Walter Reed, confirmed the hypothesis, originally put forward by the Cuban doctor Carlos Finlay, that yellow fever was spread by mosquitoes. At that point, says Alison Bashford, yellow fever containment switched from being a question of quarantine and the control of movement across national boundaries to a colonial Trojan horse. "The apparent imperative to maintain hygiene and keep out disease sets up the territorial boundaries of a nation," she explained, "but it also gives that nation an almost humanitarian license to step over its own border and into the territory of another state."

In practice, this meant that in the early years of the twentieth century, one of the central ways in which the United States pursued its colonial ambitions was through disease control, using concern about the return of yellow fever to its shores as an excuse to intervene in, influence, and eventually, in some cases, administer its southern neighbors. "In a whole lot of places like Cuba, and eventually Panama, Puerto Rico, and even Guam, the first U.S. inroads were around quarantine and infectious disease measures," Bashford told us. "Unsurprisingly, this was followed pretty quickly either by territorial acquisition of those places or by transnational agreements and other extensions of influence." Under the cover provided by quarantine, the United States was able to pursue an expansionist foreign policy by other means; later, U.S. president John F. Kennedy would employ the term's useful ambiguity to order a naval blockade of Cuba without committing to a definitive act of war against the nuclear-powered Soviet Union.

For an even more vivid example of quarantine's colonial footprint, Bashford pointed us to Africa, where European powers, concerned about the economic effects of infectious disease on their human capital, the native population, implemented quarantine cordons that have since hardened into international borders. During the European "Scramble for Africa" in the final decades of the nineteenth century, Britain, France, Belgium, Italy, Portugal, Spain, and Germany replaced the fuzzier tribal boundaries of precolonial Africa by drawing their own lines on maps. "We have been giving away mountains and rivers and lakes to each other," wrote the British prime minister Lord Salisbury in 1890, "only hindered by the small impediments that we never knew where the mountains and rivers and lakes were." Quarantine, as always, proved an effective way to make those arbitrary lines seem real.

Take the international border between Egypt and Sudan, which the British originally laid down in a dead straight line following the twenty-second parallel in 1899, at a time when they effectively controlled both countries. Today, there are just three deviations from that perfect geometry, the result of British administrative adjustments. One of them, the Wadi Halfa salient, is a small, finger-shaped area that pokes north from Sudan into Egypt along the Nile—and the site of a former quarantine station. This facility was placed strategically at the terminus of the rail line from Khartoum, at the point where goods and people transferred to steamships on the Nile to continue their journey.

The story behind the Wadi Halfa wiggle begins in 1919, when the British began construction on a monumental irrigation project south of Khartoum, damming the Blue Nile in order to grow cotton in a region known as the Gezira. To do the work, they imported more than forty-five thousand Egyptian laborers on six-month contracts, from a region in which schistosomiasis, a parasitic worm infection also known as bilharzia, was endemic. Gezira, by contrast, was disease-free. In the words of Major B. H. H. Spence of the Royal Army Medical Corps, writing in November 1924, the authorities, "recognizing that full economic value could not be obtained from a horde of diseased workmen, and fully alive at the same time to the danger that would be incurred by introducing them in their native state into the Gezira, decided to establish a quarantine at Wadi Halfa."

Twice a week, hundreds of Egyptian laborers arrived at Wadi Halfa by steamer, spending four days in quarantine before going onward to Khartoum by train. In order to keep track of each individual as they passed through the labyrinthine facility, officials painted numbers on their forearms using a silver nitrate solution, which darkened in the sun and remained indelible for several days. The unfit were isolated and sent home, and the rest were thoroughly sanitized. They were starved for twenty-four hours prior to deworming, and then catheterized and their feces collected in numbered bowls, so that their excretions could be examined for evidence of infestation. In groups, they were then stripped naked, shaved from head to toe, steam fumigated, and vaccinated against smallpox. "Did the fellah know anything of the mythology of his country he might be tempted to compare the horrors of his passage through the quarantine with those which the souls of his ancestors encountered on their journey through the portals of the underworld," concluded Spence, with what passed for sympathy. "Probably, however, his thoughts centre mainly on the much more pressing problem of when he is going to get his next meal."

Two thousand miles to the south, the Belgians were implementing a quarantine line of their own, in order to protect the Uele district, an "uncontaminated triangle" at the northeastern corner of the current Democratic Republic of the Congo that was judged to be free from sleeping sickness. This Belgian-ruled province bordered what is now the Republic of the Congo, a French possession, as well as the British-held Ugandan Protectorate, where the disease was rife: a quarter of a million Ugandans died of it between 1900 and 1920. As the historian Maryinez Lyons has pointed out, those colonial borders may have looked definitive on maps in London, Paris, and Brussels, but, on the ground, the entire region was tied together by social and economic networks—fish catching and its subsequent salting, trade routes, and kinship bonds—that its inhabitants, unsurprisingly, were keen to maintain.

Instead, the Belgians established a cordon around the Uele province, with sleeping sickness checkpoints and lazarettos making the colony's border manifest. Entire villages were cut off from their fishing grounds or trading partners, forbidden to travel across quarantine lines without a medical passport. Any African with swollen lymph glands, a characteristic symptom of the disease, was detained in a lazaretto—known locally as a "death camp"—or resettled somewhere else in the Congo, in an infected district. Maintaining a strict cordon in such a remote location was an investment, but, as in the case of Wadi Halfa, it was hardly motivated by colonial benevolence: instead, as a senior Belgian administrator wrote in 1911, "the populations of Uele seen from a simple economic point of view represent such capital that it seems to me no sacrifice to save them."

Thus, through a combination of social engineering and medical infrastructure on the ground, the Belgian Congo established its northeastern frontier. Later, Alison Bashford explained, when Uganda, the Democratic Republic of the Congo, and the Republic of the Congo became independent nations in the 1960s, these quarantine lines "offered a clear and politically useful demarcation" for their new international borders. The quarantine line around Wadi Halfa, on the other hand, is still disputed, although Egypt has all but rendered the quarrel moot: the entire salient was drowned following the construction of the Aswan dam. Nonetheless, in both cases, Bashford pointed out, colonial-era public health restrictions on mobility created their own geopolitical reality—a legacy that complicates the lives of Africans to this day.

⟷

As the nineteenth century unfolded, in spite of the terror that accompanied cholera's cyclical return to the continent and the arrival of yellow fever, quarantine began to fall out of favor. Critics claimed that it was arbitrary and, for the most part, useless, even while it also exacted an enormous economic cost in terms of lost time and trade. Admittedly, no plague outbreaks had occurred on Austrian territory following the construction of its sanitary cordon—but plague, for reasons that public health historians still debate, had largely retreated from continental Europe by the end of the 1700s. Meanwhile, as Emperor Joseph II complained, too much trade had been lost to the Dalmatian coast, where the Venetian controls were seen as less strict.

Britain, which once conducted so little trade with the East that it didn't bother to build a lazaretto, had, by the start of the nineteenth century, become the world's dominant maritime and mercantile power, its colonial possessions augmented by the first global trading empire, overseen by the Royal Navy. Most British politicians understood that the country's prosperity was now based on the free flow of goods and people around the globe, unencumbered by quarantine.

At the same time, they were aware that plenty of other countries did not share that view—or that economic model—and would happily subject British ships to retaliatory quarantine if they felt its sanitary precautions were not adequate. Mobility, and thus prosperity, was dependent on mutual trust. The country's temporary solution, following the first two European cholera outbreaks, Denis Vandervelde told us, was to outsource quarantine—to Malta. "For about ten years, Malta was the busiest quarantine station in the world," he said. "From being quite a rare piece of postal history, from late 1837, it's the commonest of all disinfection marks—you can buy a reasonable strike for twenty dollars."

Malta, Vandervelde reminded us, was at the time a British possession. The British teamed up with the French, who were known for their exemplary quarantine facilities at Marseille, to operate a lazaretto that quickly became known as the Mediterranean's most efficient. Ships that performed quarantine at Malta were automatically given free pratique for Britain or France, in a system that resembles today's border pre-clearance facilities, which smooth the path for travelers to clear the U.S. customs and immigration checkpoints in, say, Shannon Airport in Ireland, and enter the country as domestic travelers.

Such shortcuts were warmly welcomed by many. As traders, travelers, and hajj pilgrims were all too aware, quarantine was not only inconvenient but, all too frequently, both brutal and corrupt. At El Tor, on the Suez Canal, armed guards patrolled the camp on camelback, with orders to shoot escapees, but rich pilgrims frequently paid others to stand in for them at disinfection sessions, or shortened their detention with a well-placed bribe. Meanwhile, at the international level, quarantine was often abused for political gain. As Mark Harrison, a historian of medicine, has noted, during a 1770 plague outbreak, Prussia strategically established a sanitary cordon that encroached on Polish territory, "its ostensibly defensive nature masking predatory intentions." Quarantine officials everywhere shamelessly opened and read official dispatches under the pretext of disinfection.

One British diplomat complained that "in the name of public health," the Russians "had introduced a system of universal police and espionage." Later, in 1823, French troops mustered at the Spanish border, forming a cordon sanitaire to stop the spread of both yellow fever and Spanish liberalism; the following year, they crossed their own line, invading to help restore the Bourbon monarchy to the Spanish throne.

Some of quarantine's failures have been fatal: Vandervelde showed us the faked paperwork for a Genoese ship transporting wool, linen, and tobacco from the Peloponnese, where plague was rampant, to Sicily. By obscuring his port of origin and explaining the loss of one plague-stricken crew member by saying he had fallen overboard, the ship's captain was allowed to land his cargo; a few months later, an estimated sixteen thousand Sicilians had died from the Black Death. For the most part, however, quarantine presented an almost unbearably frustrating, economically painful delay. Witness a scorched letter from a ship's captain held in the Livorno lazaretto, dated May 23, 1788: "Forty days since any suspect vessel arrived . . . a disgrace! There is nothing in my papers to say how long the quarantine will last," Vandervelde translated. "I am losing because there is no other ship loading for Genoa or Marseille and I could have filled my space, but other ships are now arriving."

By the mid-1800s, agreement that quarantine was in need of reform, at the very least, was widespread in Europe. In the 1820s and 1830s, the continent had embarked on a burst of postal treaty-making, regularizing the international delivery of mail. Why not also standardize quarantine? With the widespread revolutions of the 1840s out of the way, the French gathered twelve European powers for the first International Sanitary Conference, under the motto "maximum protection with minimum restriction."

The delegates failed to reach a binding agreement—only Sardinia ratified the resulting convention—but it was followed by ten more International Sanitary Conferences over the next five decades. That initial acknowledgment of mutual interdependence paved the way for dozens of world-organizing agreements, among them the International Telegraphic Union (1865), the Universal Postal Union (1874), and the International Bureau of Weights and Measures (1875). More directly, the International Sanitary Conferences are the direct ancestor of today's World Health Organization.

"Communicable disease and quarantine always put the idea of cooperative international governance on the table, and often drew lines of communication, transport, and movement into discussion," Alison Bashford told us. "The question of how to manage infectious disease"—and thus the International Sanitary Conferences convened to resolve it—"is foundational for the League of Nations Health Organization, and then the World Health Organization." Quarantine logic—originally used to manage national borders, and then to define otherwise illusory cultural regions—had evolved into the infrastructure of global governance that still regulates trade, commerce, and human health in the twenty-first century. Quarantine restrictions, we came to realize, lie at the root of most global institutions and frameworks, preserved like a fly in bureaucratic amber.

⟷

The underlying goal of the International Sanitary Conferences was to find a way to secure borders more flexibly. Could the inflexible inconvenience of quarantine be replaced by a system that allowed the blessings of trade to flow unhindered while creating a barrier against loathsome diseases and their carriers?

When the first meeting was held, in 1851, most European countries believed quarantine was essential, albeit flawed. Britain, however, was so economically wedded to free trade that it had begun to devise measures other than quarantine to prevent the importation of infectious disease. In the 1850s and '60s, the country's initial solution of simply outsourcing quarantine to Malta was replaced by a combination of measures that came to be known as "the English preventive system": a multifaceted approach that involved epidemiological intelligence and investments in public sanitation, designed to catch and isolate sick people on and even after arrival, rather than detaining everyone at the border.

There were no cholera outbreaks in England after 1866, and the country also boasted falling death rates from tuberculosis. Gradually the rest of Europe took note: the historian Anne Hardy writes that, in the 1890s, the English system was pronounced by both French and Finnish experts to be "the most complete and precise in the civilized world." By the end of the century, thanks to the rise of global trade and a growing understanding of the science of disease transmission, the rest of Europe was on board.

This neo-quarantine, pioneered by the British and eventually adopted by the rest of the world as the basis of global health, still operated by controlling mobility (of people and thus their germs). It simply replaced the physical barrier of lazarettos and cordons sanitaires with a selective, surveillance-based one—one that relied on data rather than buildings. As the historian John C. Torpey has written, this shift, from fixing people in space with architecture to tracking their movements and contacts, requires "an identification revolution"—the universal implementation of "techniques for uniquely and unambiguously identifying each and every person on the face of the globe, from birth to death," as well as the construction of bureaucracies "to scrutinize persons and documents" in order to verify those identities.

PEr lettere di diuerſi Magiſtrati di Sanità eßendoſi inteſo che per guardarſi dalla Peſte che fà gran ſtrage in alcune Città della Germania, & in diuerſi altri luoghi già banditi, hanno ſtabilito d' introdur' l'vſo delle Fedi di Sanità così, per le perſone come per le robbe gli Signori Conſeruatori della Sanità con participatione dell' Eminentiſsimo, e Reuerendiſsimo Sig. Cardinale Mareſcotti Legato, hauendo deliberato di porre in vſo dette Fedi in queſta Città, e Diſtretto a finche à queſti Sudditi, e Viandanti non ſia vietato l' ingreſſo in altri Stati, fanno ſapere a tutti che nell' Vffitio del Comune ſarà continuamente perſona deputata à queſt' effetto che le darà Gratis per ogni Città, e luogo doue vorrãno andare, e confermarà quelle che nel tranſito preſentaranno li paſſagieri che verranno da altri luoghi non ſoſpetti &c.

Dat. nell'Vffitio della Sanità li 17. Ottobre 1679. Pub. 18. det.

Li Conſeruatori di Sanità.

Alfonſo Gioia Canc.

In Ferrara, nella Stampa Camerale.

A poster from Ferrara, Italy, printed in 1679, announces that health passports will be required for all travelers. (*Courtesy of the Wellcome Collection*)

Today, we know these all-encompassing techniques and bureaucracies as the "passport" and "passport control," but the earliest such documents were health passports—an innovation that, like quarantine, dates back to the plague era in Italy. From the 1500s, local authorities would issue these formal printed documents, known in Italian as *fedi di sanitá*, to travelers hoping to avoid quarantine at their destination. As "a bit of a breather" between showing us his collections of quarantine letters and disinfected mail, Denis Vandervelde allowed us to inspect some of the earliest health passports in his collection. "The reason I've selected these is that, until about 1700, paper was very expensive and therefore they use very small sheets," he said. "As time went on and paper became cheaper, the passports got bigger and bigger and bigger and ended up enormous—so these obviously are easier to show."

From the 1500s, health passports were required for travel in much of Europe, and usually issued without charge; like today's passports, they requested free and unhindered passage for their bearer, although that plea was not always granted. In 1636, William Harvey, the British physician known for describing the circulation of blood, was detained in quarantine at the lazaretto in Treviso on his way from London to Venice, despite presenting a valid health passport. "I have receyved heare a very unjust affront," he wrote to his host in Venice, complaining that he was being wrongly held in "a very nasty roome" and provided with an "ill diett." The whole episode had, he wrote, brought on a bout of sciatica "that much discorageth me, and maketh me lame."

Later, when vaccinations had been developed for certain diseases, including smallpox, the use of a paper-based health passport was often replaced with the inspection of a bodily one. "You could only move from what was understood to be an infected zone into a healthy zone if you had a vaccination scar," Alison Bashford told us, describing the way that, for example, during an 1881 epidemic of smallpox in Sydney,

Australia, passengers on the train bound for Melbourne were stopped at the state border and sent back unless they could display a distinct, circular, upper-arm cicatrix. "In these developing and increasingly global and governmental systems of surveillance and of identity documentation," she explained, "the vaccine scar was a significant corporeal identity document in and of itself."

⟷

Like quarantine, after the 1950s, "disinfected mail more or less disappears," Denis Vandervelde told us. The most modern item in his collection dates to 1972, when, just as the World Health Organization was preparing to declare smallpox eradicated, a Yugoslav guest worker in Hanover, Germany, was taken to the hospital and quickly isolated, suffering with what local doctors described as "a fulminating pox." He had been in Germany for only two weeks, but smallpox is highly contagious: it was essential to track down all his contacts, as many of them might also be infected. There was just one problem. "He was a very handsome young man," said Vandervelde, "and when they put his picture in the local newspaper and asked for anybody who'd seen this man to come forward, 283 girls claimed to have met him."

The authorities dutifully rounded up all 283 young women, housing them in village halls and Scout huts for a period of quarantine. "Some of them had parents who didn't have telephones," explained Vandervelde. "So they had to allow them to post mail." Smallpox, as it happens, is one of the very few pathogens whose transmission through the mail has actually occurred—Vandervelde told us that, during the American Civil War, there were six authenticated cases of wives or girlfriends who, on receipt of a letter from their smallpox-stricken beau at the front, "kissed it or put it into her bosom and later went down with smallpox." With their old rastels and tongs consigned to the museum, German authorities decided to disinfect the girls' letters by wrapping them in muslin and ironing them three times at the highest setting.

"Here's where the story becomes murky," continued Vandervelde, clasping his hands together in evident delight. Those letters were stamped to say they'd been treated against smallpox, but in blue or purple ink rather than the more usual black, which immediately raised Vandervelde's suspicions. After some digging, he concluded that one of the local doctors was also a stamp collector, and had seen this outbreak as his opportunity to make postal history. He designed and made up special stamps, which he offered to the post offices. This makes the cancellations only "semiofficial," according to Vandervelde, who owns a handful of them. "They're official in the sense that they were accepted by authorities, but they weren't issued centrally," he explained. In postal history competitions, judges look askance on material that is deemed to be "contrived"—so Vandervelde has never displayed his German smallpox disinfection materials for fear they may be dismissed as merely an elaborate artifice.

Even more recently, following the 2001 anthrax attacks, the U.S. Postal Service determined that letters and parcels sent to zip codes beginning in 202, 203, 204, and 205, which serve federal government agencies in Washington, D.C., should be treated. A company that irradiated food to extend its shelf life won the contract, and, although the USPS has declined to comment, the agency's website says that mail destined for those zip codes is still forwarded to New Jersey, where it is put on a conveyor belt and passed under a high-energy beam of ionizing radiation to kill bacteria and viruses. The letters and packages are then "aired out" for a while, before being forwarded to their destinations. The paper is left slightly faded and somewhat crispy, but sterile—and, at least sometimes, stamped as such. "Those cachets typically go for about twenty-five dollars," said Vandervelde. "I think even that's overpriced for something that recent."

In the early days of the COVID-19 pandemic, as public health authorities resorted to the medieval technology of quarantine, they also wondered whether they ought to resurrect mail disinfection. Early studies indicated that the virus could survive for twenty-four hours on cardboard surfaces, perhaps longer on paper. In February 2020, China's central bank began quarantining the country's cash, collecting banknotes from Hubei, the worst-hit province, then baking them at a high temperature or bathing them in ultraviolet rays. The newly laundered cash was then kept in isolation for seven to fourteen days before being rereleased. A few weeks later, the U.S. Federal Reserve began quarantining dollar bills repatriated from Asia, holding them for seven to ten days before allowing them to reenter the domestic financial system. As infections in America soared, dozens of countries refused to accept mail sent from the United States: in April, a birthday gift we sent to a nephew in Bermuda was returned to us with a sticker saying "Mail Service Suspended."

Just as in centuries past, in these times of quarantine, new health borders are being drawn, and new forms of disease-containment surveillance and control are being prototyped. In 2020, so-called travel bubbles or corona corridors—a means by which British holidaymakers traveling to Spain could sidestep quarantine, for example—set the stage for novel geopolitical alignments in a formerly borderless Europe. Thermal screening at airports became the new vaccination-scar inspection: a corporeal license to cross borders. Health officials even proposed "COVID passports" to identify those who, either by virtue of having recovered from the virus or having received a vaccination, were now immune and able to resume normal life.

In China, a new system was rolled out in cities across the nation in response to the novel coronavirus. Called Alipay Health Code, a QR symbol linked to a popular mobile payment system classified individuals according to infection status: a modern-day health passport, reporting your location data and identity to the government while redefining constraints on your mobility in real time. A square green bitmap on your phone would open subway turnstiles, lift freeway tollgates, and allow entry to public markets, restaurants, shops, and banks. Renting a shared bike in an area with an elevated infection rate might be enough to trigger a yellow code, shutting off access to the city for seven days of quarantine. If you code red at a checkpoint, indicating either a confirmed infection or close contact with someone known to have COVID-19, the police will be called. It is not hard to imagine authorities finding such a tool useful long after this particular pandemic has ended. Immediately, individuals began hacking the system, showing a screenshot of a green code after theirs had turned yellow or borrowing a friend's phone to get around, just as the Genoese ship's captain lied to dodge quarantine in Sicily during the Black Death, and the rich paid stand-ins to undergo disinfection at El Tor.

We emerged from Denis Vandervelde's living room blinking in the amber haze of late-afternoon sunlight, the rows of tidy mock-Tudor semidetached houses casting long shadows across his street. As we made our way to the nearby Tube station, we remembered that the land just to our northwest had once been the property of the Knights Hospitaller, who first built the honey-hued sandstone lazaretto on the island of Malta, and whose successors, the Order of Malta, once again swung into action during COVID-19, setting up quarantine facilities for refugees in Germany and a mental health hotline for those struggling during lockdown in Italy.

As the enduring persistence of both vampire myths and African border disputes demonstrates, the temporary infrastructure and controls on mobility put in place during a pandemic often harden into permanent borders, bureaucracy, and, too often, inequities: the "new normal." Today's world is structured by the ghosts of quarantines past; tomorrow's is now taking shape all around us, built out of expedience and fear.

4

An Extraordinary Power

By the time a person dies from Ebola—and, in some outbreaks, nine out of ten infected individuals do—their face has often assumed a peculiar, hauntingly blank cast, with sunken eyes and fixed features. This zombie-like death mask is a terrifying but characteristic sign of the disease, in both monkeys and humans. In part, it happens because the virus is capable of attacking the central nervous system, damaging the parts of the brain that control facial expression in addition to causing confusion, seizures, and even psychosis. Ebola also preferentially consumes connective tissue, leaving the upper layer of the skin to float on a layer of liquified collagen. In the most severe cases, as the virus replicates exponentially inside cells, triggering rampant inflammation and necrosis, the corpse will liquify within hours of death, discrete organs reduced to jelly.

Over little more than a week, a truly unlucky individual will have progressed from "dry" symptoms—fever, aches, and fatigue—to the "wet" stage of the disease, in which standard-issue diarrhea and vomiting are sometimes accompanied by uncontrolled bleeding from every orifice and membrane: blood weeping from glassy, reddened eyeballs, oozing from inflamed gums and previously healed scars, and leaking from the nose, anus, and vagina.

Ebola is thought to have first emerged in humans in 1976, making it impossible that Edgar Allan Poe could have known about it when he wrote "The Masque of the Red Death" in 1842; nonetheless, doctors have proposed naming one of the viral strains Ebola-Poe, "in honor of the creative genius that imagined the horrors of hemorrhagic fever long before the disease was recognized." "Blood was its avatar and its seal—the redness and the horror of blood," Poe wrote, describing the existential terror that his Ebola-like contagion provoked. "The scarlet stains upon the body and especially upon the face of the victim, were the pest ban which shut him out from the aid and from the sympathy of his fellow-men."

Until 2019, there was no vaccine for Ebola, and no treatment other than supportive care. Fear has been the ubiquitous and not unreasonable response to the disease, with the unfortunate result that Ebola sufferers and their families have often become Poe's pariahs. In 2014, when the largest epidemic to date spread through Sierra Leone, Guinea, and Liberia, killing at least 11,300 people, many were terrified that they might catch the disease, afraid of the Ebola Treatment Units from which so few returned, suspicious of their neighbors, and wary of survivors.

At the height of the outbreak, one in seven people in Sierra Leone were under quarantine. In next-door Liberia, authorities quarantined entire villages, at one point building a makeshift cordon from razor wire and chipboard to wall off West Point, one of the most densely populated slums in the capital city of Monrovia. A few days earlier, teenagers armed with machetes and slingshots had staged a raid on a West Point Ebola Treatment Unit (ETU), "freeing" twenty infected individuals who promptly disappeared into the neighborhood.

In a series of interviews compiled by CDC medical anthropologists, Monrovians confessed that they preferred to hide sick family members rather than risk their removal to an ETU. "Your person going in the hospital, they don't come back," one explained. Normally, a community leader said, neighbors would support one another in case of sickness; but, in the case of Ebola, they wouldn't go anywhere near an infected person's house, and, worse, were likely to report the case to the government's disease hotline, thus ensuring the Ebola-stricken individual would end up in a dreaded ETU. Sometimes even family members would abandon a sick relative, skipping town to escape both infection and the stigma associated with the disease. "We have an uncle, he had family around but they all ran away from there," an inhabitant of West Point told the CDC. "Nobody bring him food, no water—everybody hear him screaming, but everybody afraid."

Those fortunate enough to survive were also shunned. A Guinean doctor who was one of just thirty people in the country to recover from Ebola told Doctors Without Borders that the stigmatization "is worse than the fever." No one would shake his hand, eat with him, or even remain in the same room as him. "Now, everywhere in my neighborhood, all the looks bore into me like I'm the plague," he said. By the end of the outbreak, thousands of children who had been orphaned by the virus were also turned away by their remaining relatives; a volunteer nurse said she'd seen children who survived Ebola be rejected by their own parents.

The terror proved more transmissible than the virus, with Americans exposed via their screens to newspaper articles declaring the Ebola epidemic "totally out of control," while cable networks showed apocalyptic footage of corpses left to rot in the streets. Republicans quickly saw a political opportunity in the Obama administration's refusal to ban travel from West African countries. "You don't want us to panic?" the Fox News host Jeanine Pirro said on air.

"How about I don't want us to die!" The Republican representative Mike Kelly of Pennsylvania told Newsmax TV viewers, "The government needs to stop acting as if it's absurd for people to fear a virus that liquefies their internal organs." Donald Trump—at the time, still just a reality-TV veteran and bankruptcy-prone real estate developer—shared his views on his favorite outlet, Twitter, declaring, "The U.S. must immediately stop all flights from EBOLA infected countries or the plague will start and spread inside our 'borders.'"

The disease itself initially arrived in the United States in September 2014, in the form of Thomas Eric Duncan, a Liberian man visiting family in Dallas. He died eight days later, in the isolation unit at Texas Health Presbyterian Hospital, but not before infecting two nurses—one of whom had flown to Cleveland for a wedding in the meantime—leading to the quarantine of dozens of Americans who had come into contact with them.

Hundreds more, whose mistake was merely to have been in the same city—even on the same continent—were asked to stay home from work or school as Ebola hysteria swept the nation. An elementary school teacher in rural Maine was forced to take a twenty-one-day leave of absence after concerned parents heard that she had attended a conference in Dallas, a full ten miles away from the hospital where Duncan lay. In Mississippi, middle school students were kept home following news that their principal had recently returned from a family funeral in Zambia—three thousand miles away from the outbreak's epicenter. One particularly unfortunate woman threw up as her American Airlines flight departed Dallas–Fort Worth; she was promptly locked in the bathroom by the crew for the duration of her journey, just in case.

A couple of weeks later, in New York City, Craig Spencer,

a doctor who had recently returned from treating patients in Guinea, tested positive for the disease and was placed in isolation at Bellevue Hospital—but not before he had traveled by subway, strolled along the High Line, visited a bowling alley, and taken an Uber. Media pundits were outraged: online chatter quickly condemned his behavior as selfish, reckless, and potentially lethal. On Fox News, Megyn Kelly labeled Spencer "irresponsible," random commenters suggested he should be liable for manslaughter, and, of course, Trump weighed in on Twitter: "The U.S. cannot allow EBOLA infected people back. People that go to far away places to help out are great-but must suffer the consequences!" Several state governors, including Chris Christie of New Jersey and Andrew Cuomo of New York, promptly instated mandatory twenty-one-day quarantines for anyone arriving from West Africa.

Enter Kaci Hickox, "the Ebola nurse." When Hickox landed at Newark Liberty International Airport on October 24, 2014, she had spent the past month as a medical team leader with Doctors Without Borders, based in Bo, Sierra Leone's second-largest city. A forty-bed Ebola Treatment Unit had opened just a few weeks before she arrived, and it was immediately overwhelmed; Hickox and other volunteers worked long hours, struggling to keep up with a seemingly endless flow of new patients.

To protect herself from virus-laden blood and other bodily fluids, Hickox had to treat patients while wearing multiple layers of personal protective equipment: an impermeable Tyvek chemical suit over scrubs, accessorized with rubber boots, two layers of heavy-duty gloves, a mask, goggles, hood, and, finally, an apron to cover the chemical suit's zipper. As Luigi Bertinato helped us discover in Venice, putting this outfit on correctly takes even a seasoned professional at least ten minutes. Inside the unit, conditions were so hot and humid that sweat would immediately saturate Hickox's mask and start to pool in her boots. Doctors Without Borders' guidelines allowed staff to work in full PPE for only forty minutes at a time, by which time Hickox was often squelching around in her own sweat with every step.

The conditions were tough, but the emotional toll was even more draining. On one of Hickox's first days in the unit, a teenage girl came in with Ebola symptoms; during the intake process, it emerged that seventeen of the girl's family members had died of the disease over the past three months. On her last night before flying home, Hickox was called in at midnight because a ten-year-old girl was having seizures. Hickox carefully coaxed crushed Tylenol and antiseizure drugs into her mouth in between spasms. The little girl died a few hours later, alone but for Hickox.

On Wednesday, October 22, Hickox began her long journey home to Maine. An initial short hop to Freetown, Sierra Leone's capital city, was followed by a seven-hour flight to Brussels, where she was debriefed by the Doctors Without Borders operations team. From there, she flew to the United States, arriving at the customs and immigration checkpoint at Newark's Liberty airport at lunchtime on Friday afternoon. When Hickox told the customs officer that she had traveled from West Africa, he put on gloves and, in accordance with the CDC's enhanced screening procedures, escorted her through a pair of steel doors to the quarantine station, depositing her in a tiny, windowless examination room equipped with a hospital bed and a pair of wheeled stools.

Over the next few hours, a series of people wearing different degrees of protective gear, from a face mask and gloves to a gun belt under a Tyvek suit, questioned Hickox about her work in Sierra Leone. One took her temperature using an infrared forehead thermometer. Finally, at about 7:00 p.m., Hickox was taken to University Hospital, Newark, with an eight-car police escort, sirens blaring, and deposited inside an isolation tent set up in the middle of an unfinished floor of the hospital's newest building.

Despite having no Wi-Fi and limited cell reception, Hickox managed to text her partner, family, and friends. Her partner's uncle immediately suggested she get a lawyer, and the boyfriend of her roommate from Johns Hopkins passed along the number of the attorney Norman Siegel, former director of the New York chapter of the American Civil Liberties Union. A former colleague from the CDC who had gone on to work in journalism helped her write an op-ed for the next day's edition of *The Dallas Morning News*, arguing that her quarantine was illegitimate and that the United States should "treat returning health care workers with dignity and humanity," rather than making them "feel like criminals and prisoners." "She felt like you would treat a dog better than she's been treated," Hickox's mother told *The New York Times*.

That weekend, Hickox's vocal efforts to be released from quarantine led on every news channel. A handful of sympathetic voices, including Dr. Anthony Fauci, director of the National Institute of Allergy and Infectious Disease, chimed in, criticizing her quarantine as unnecessary and unhelpful. Overall, however, most Americans seemed to think that quarantine was a good idea—"better safe than sorry," as several politicians and journalists put it—and that Hickox's reluctance to stay in isolation for three weeks after returning from the hot zone was selfish at best, if not malicious. "After what she saw in Africa, I find it deplorable that she would not do what was advised to ensure everyone's safety," said one commenter. "What is a few weeks in the overall scheme of a lifetime?"

On the campaign trail for the upcoming midterm elections, Governor Chris Christie seemed to relish the fight. "I believe that folks who are willing to volunteer also understand that it's in their interest and the public health interest to have a twenty-one-day quarantine period thereafter if they've been directly exposed to people with the virus," he told reporters. "I will take whatever steps are necessary to protect the public health of the people of New Jersey, and if someone wants to sue me over it, they can."

⟷

"I have *a lot* of personal thoughts about the misunderstanding and misinterpretation of what this tool is," Dr. Martin Cetron warned us the first time we spoke. "Quarantine is something that I'm pretty close to and pretty passionate about." A youthful sixty-one-year-old with a cheery, round face and a penchant for aphorisms, Cetron is both a medical doctor and a retired captain in the U.S. Public Health Service, an agency founded more than two hundred years ago as the Marine Hospital Service, charged with preventing sailors and immigrants from importing disease into the nascent United States. We met him at his offices inside the shimmering, green-tinted glass blocks of the CDC campus in Atlanta, Georgia, where he goes by yet another title: director of the Division of Global Migration and Quarantine.

"When I came to the division in '96, I proposed an alternative name," Cetron confessed. "I tried to get rid of the word *quarantine*." Over the years, however, as his proposal slowly made its way up the chain of command, Cetron had a change of heart: *quarantine* could and should remain, but it desperately needed to be reformed—and rebranded. "There's so much—*so much*—loaded in that word," he said, "much of it grounded in its misuse and abuse in history."

Today, Cetron is the closest quarantine has to a poster

child. As we reported this book, Cetron popped up at every turn, keynoting conferences on reducing the spread of communicable diseases at airports, participating in panel discussions on the potential for big data to help contain pandemics, being mobbed by attendees with questions about quarantine at postworkshop receptions. "Every major global epidemic of disease is followed by an epidemic of fear and an epidemic of stigma," he would say each time, urging empathy and moderation from scientists and Transportation Security Administration screeners alike.

Cetron, who also sits on a handful of WHO Expert Committees, has led the U.S. containment efforts in response to most of the major outbreaks of the twenty-first century, including the anthrax attacks of 2001, SARS in 2003, swine flu, MERS, and, of course, the 2014 Ebola outbreak. It was based on his recommendation that the Obama administration resisted calls to close U.S. borders to flights from West Africa; instead, Cetron worked with Liberian, Guinean, and Sierra Leonean authorities to set up exit screening at their airports and ports, in a largely successful effort to catch exposed and infected individuals before they boarded an airplane. (Thomas Duncan being the unfortunate exception. "There was denial in the patient," said Cetron diplomatically.) In 2015, at the request of Margaret Chan, the former director-general of the World Health Organization, Cetron helped talk panicked athletes, scientists, and bioethicists down from a mass campaign to cancel the Rio de Janeiro Olympics by showing that the risk of increased Zika transmission was, in reality, vanishingly small.

Cetron is thus all too aware that public perception of a disease's threat is frequently incommensurate with reality, and that the natural human dread of the death and disruption that accompany pandemics can easily escalate into a hysterical impetus to do something—anything!—to protect ourselves. "One of the problems is that quarantine is used as a political tool in an overreaction to fear," said Cetron. "And that has given it a really bad name."

To make matters worse, the state's quarantine authority is an extraordinary power: the complete inversion of the presumption of innocence that otherwise underlies Anglo-Saxon legal thought. As Australian legislators put it, while establishing the newly unified continent's sanitary policy in 1884, quarantine "differs from a measure of criminal police in this respect: That it assumes every person to be capable of spreading disease until he has proven his incapacity; whereas the law assumes moral innocence until guilt is proved."

That reversal, explained Krista Maglen, an Australian historian of medicine, makes quarantine one of the singular instances in today's liberal democracies in which a state can detain someone without having to demonstrate their guilt—merely their potential for causing future harm. By definition, the quarantined are not known to be a certain danger. Instead, they are suspected of posing a risk—an assessment that can easily be colored by bias, unconscious or otherwise. "It's that idea of risk that can be very easily manipulated," Maglen said. "It's a very loose and dangerous term."

Humans are notoriously bad at estimating risk at the best of times, but the outbreak of a deadly, unfamiliar infectious disease triggers a phenomenon that the legal scholar Cass Sunstein has dubbed "probability neglect." This cognitive blind spot occurs when a possible outcome—whether it be winning the lottery or catching Ebola—gets a grip on an individual's emotions, to the extent that their thoughts focus exclusively on the horror (or delight) of the outcome itself, at the expense of considering the likelihood that it will actually happen.

These kind of emotion-driven risk assessments are typically instinctive and instant—gut reactions, governed by preexisting assumptions, many of which are based on discriminatory stereotypes. Scientific risk is arrived at by factoring in the nuances of the pathogen in question, an individual's personal exposure, susceptibility, and behavior, as well as environmental conditions. Perceived risk, on the other hand, might simply rely on the—often racially motivated—belief that the place that a person has come from seems somehow dirty or diseased.

Maglen's home country, Australia, provides a clear example of this kind of bias-driven misuse of quarantine. As the British colony moved toward independence in the late 1800s, it began to frame its geographic neighbors in Asia as "the Yellow Peril"—an existential threat of contamination against which the virginal purity of this isolated island nation had to be defended at all costs. The first director-general of the Australian Department of Health, John Cumpston, spelled out the objectives of the national quarantine policy in unambiguous terms: the creation of a white Australia. The purpose of quarantine, he wrote, is "the keeping of our continent free from certain deadly diseases," but also, and equally important, "the strict prohibition against the entrance into our country of certain races of aliens whose uncleanly customs and absolute lack of sanitary conscience form a standing menace to the health of any community." (Australia is literally depicted as a white continent, surrounded by dark-shaded countries in which smallpox was endemic, in a map illustrating a 1912 government report on quarantine.)

Unsurprisingly, the United States—like Australia, a country founded by immigrants who, once the skin color and religion of those arrivals started to shift, decided to roll up the welcome mat—has also been guilty of conflating fear of disease and xenophobia. The first book that Marty Cetron read when he began his new job, he told us, was the medical historian Howard Markel's *Quarantine!*, an account of the ethnic scapegoating that occurred during New York City's 1892 typhus outbreak. Markel shows that, in the second half of the 1800s and in the early 1900s, the demographics of undesirability varied by port of entry. In New York City, for example, quarantine policies targeted first Irish immigrants and then Eastern European and Russian Jews. The public health rationale was that they were likely to carry either cholera or typhus, the extremely contagious and much feared louse-borne disease whose victims likely included the British quarantine reformer John Howard and whose sufferers characteristically developed "the repulsive smell of rotting straw."

In reality, both are diseases of poverty, to which people are rendered susceptible by overcrowding and bad sanitation, rather than nationality or religion. Cases of typhus were also found among native-born New Yorkers, as well as Scandinavian and German-born immigrants. All too often, Eastern European Jews and the Irish caught typhus *after* they had settled in the crowded tenements of the Lower East Side—or, heartbreakingly, while waiting to enter the country, in the filthy conditions of quarantine, where hundreds of hopeful immigrants died and were buried, unrecorded, in mass graves.

One gray morning in October, we took a ferry from the tip of Manhattan, past the Statue of Liberty, to the Roman Catholic Church of St. Peter on Staten Island, for a funeral service to honor some of those unknown quarantine dead.

As we waited outside the church next to representatives of the Friends of Abandoned Cemeteries of Staten Island, who had organized the ceremony; members of the Ancient Order of Hibernians, dressed in cable-knit Aran sweaters; and local reporters, the sky darkened and it began to drizzle. The mournful drone of bagpipes playing "Going Home" drifted down the street, as a half dozen men and women in kilts, tasseled socks, and berets slowly followed a hearse containing two caskets. Inside were the partial remains of thirty-seven immigrants: the adults in one and nine children under the age of twelve in the other.

During the service, we learned what little is known about their stories. They were likely Irish, among the 1.5 million who, rendered destitute by a different disease, the fungus-like potato blight, immigrated to the United States between 1845 and 1855. Their first—and, in the case of these thirty-seven unfortunates, only—stop was the New York Marine Hospital. Built in 1799, the hospital was simply known as "the Quarantine," and it occupied a thirty-acre site surrounded by six-foot-tall brick walls just a few yards south of where the Staten Island Ferry docks today. By the late 1840s, the facility held as many as 1,500 people at a time, with a large staff observing the seemingly healthy majority, treating the sick, and burying the dead. In a busy year, accounts record that the Quarantine got through 108,010 pounds of bread, 1,334 pounds of coffee, 23 gallons of brandy, 1,300 leeches, and 556 coffins.

No such logbooks were kept of either the healthy or the sick. Those who died, including the unknown thirty-seven we had gathered to honor that day, were stripped of their clothing and belongings—which were then burned to prevent contagion—and buried in rough pine coffins stacked in trench graves, with no marker.

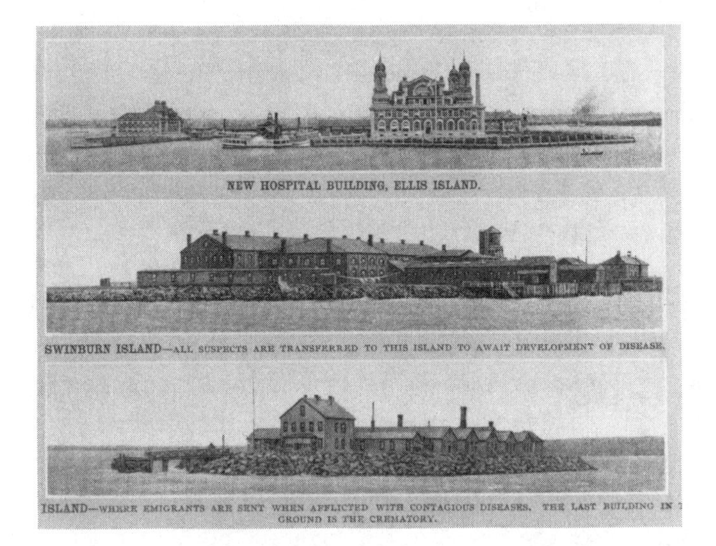

A photo collage from 1902 shows three quarantine and isolation facilities in the New York City harbor: Ellis, Swinburn [*sic*], and Hoffman Islands. Swinburne and Hoffman are both artificial islands, adding to their mystique. (*Courtesy of the U.S. Library of Congress*)

In 1858, the Quarantine was burned to the ground by a group of Staten Islanders who did not appreciate living next door to hundreds of impoverished and potentially infectious foreigners. They had been petitioning the state to close the facility for decades, complaining of both outbreaks and the effect on their property values, and finally took matters into their own hands. The building was initially replaced by a fleet of quarantine ships, but, when that proved too expensive to maintain, quarantine changed the shape of the harbor itself, as city officials piled landfill atop shoals to build two new islands—Hoffman for observation and disinfection, and Swinburne for the visibly sick. Today, both islands are abandoned and off-limits: from a boat we saw seabirds, basking seals, and what looked like the remains of a crematorium chimney on Swinburne. Ironically, given the impact of European diseases on indigenous Americans, the marble statue of Christopher Columbus that still sits atop a column in the middle of Columbus Circle in Manhattan was briefly quarantined on Hoffman, alongside the Italian immigrants who had accompanied it on its journey to New York.

Back on Staten Island, houses were built atop the Quarantine cemetery, then demolished and replaced with a municipal parking lot in the 1950s. In 2006, the city conducted a preparatory archaeological survey of the site, prior to starting construction on a new courthouse. The majority of the cemetery's inhabitants were left in place, but the jumbled remains of these thirty-seven were excavated in the process and sent to the forensic archaeologist Tom Amorosi's lab. Amorosi determined their sex, age range, and, from their compound fractures and arthritic joints, that "these people were used as mules."

With only these details to go on, we mumbled prayers, bowed our heads for the sprinkling of holy water, and, with a bagpipe salute, bid these nameless victims of quarantine farewell. They were finally interred, a century and a half after their aborted arrival, in a vault beneath the new courthouse.

Of course, inhumane conditions at the Staten Island Quarantine, and even temporary detention on the harbor's bleak quarantine islands, could only delay—not prevent—the arrival of hundreds of thousands of immigrants in New York City. But a more lasting effect of ethnic quarantine is the seemingly scientific endorsement it gives to existing prejudices. "Quarantine is very much about reaffirming models and stereotypes within the community, to create a feeling that 'everybody knows that people from a particular country or region are dangerous, because look, the government has to quarantine everybody from there,'" Maglen told us.

Incredibly, even the charities that supplied aid to New York City's newest immigrants referred to them as "human maggots."

On the West Coast, it was the Chinese who were the primary target of anti-immigrant prejudice. During the California gold rush, San Francisco had become the port of entry for hundreds of thousands of Chinese people, fifteen thousand of whom were crowded into a dozen square blocks of decrepit housing stock just north of an otherwise prime location, Union Square. The Chinese Exclusion Act of 1882 had prevented further immigration, but San Franciscans were keen to remove the "blight" that was Chinatown, and, ideally, to raze and redevelop those blocks. They tried to use quarantine as a tool to do so: when a forty-one-year-old Chinese lumber salesman named Wong Chut King was found to have died of the plague in March 1900, authorities responded by sealing off the entire district with rope, fence posts, and barbed wire. Policemen stood guard twenty-four hours a day, preventing people or goods from crossing the line, and the city's streetcars passed through without opening their doors.

Trapped, San Francisco's Chinese inhabitants were hungry, anxious about what would happen to their jobs with white employers elsewhere in the city, terrified that the authorities' next step would be to imprison them or burn down their houses, and angry at the unfairness of it all. Thousands expressed their frustration in the form of riots, which were quickly subdued by baton-wielding police. Jew Ho, a Chinese grocer on Stockton Street, was so enraged by the way the quarantine line zigzagged to exclude the white-owned businesses next door to his that he took the San Francisco Board of Health to court, where the city's ambitions were unexpectedly thwarted. In his ruling, Judge William Morrow noted that the quarantine had been applied with "an evil eye and an unequal hand," because it was "boldly directed against the Asiatic or Mongolian race as a class, without regard to the previous condition, habits, exposure to disease, or residence of the individual," in direct contradiction of the equal protections guaranteed by the Constitution.

But quarantine in the United States hasn't been abused only in an attempt to obstruct the passage of undesirable immigrants at the border and stigmatize those who have already arrived. One of the least-known quarantines in U.S. history is the Orwellian-sounding "American Plan," under which thousands of low-income women were detained on suspicion of spreading venereal diseases. The Yale law student Scott Stern spent years digging through archives to write the first book-length account of the program, published as *The Trials of Nina McCall* in 2018.

The Plan was introduced in 1917, as the United States

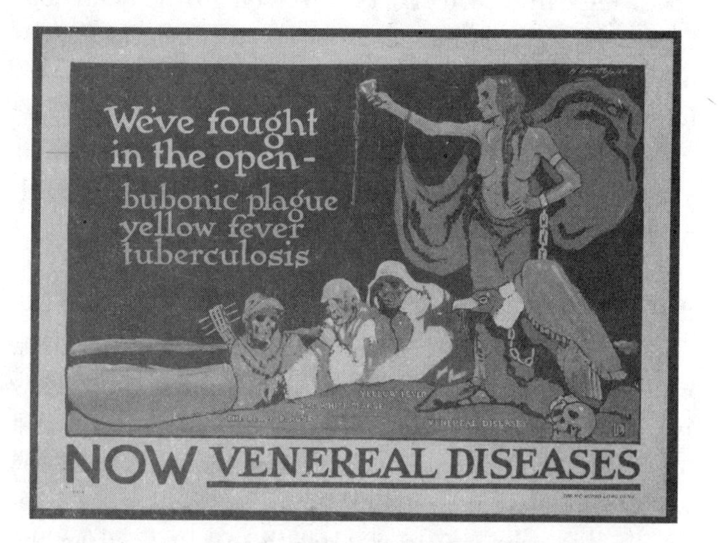

A poster by the artist H. Dewitt Welsh, ca. 1918, warns against the perils—both medical and moral—of venereal diseases. (*Courtesy of the U.S. Library of Congress*)

entered World War I, to protect the nation's young male fighting force from sexually transmitted infections by empowering local officials to quarantine and medically inspect any woman whom they "reasonably suspected" of having syphilis or gonorrhea. In reality, Stern shows, the program was the establishment's misogynistic and paternalistic response to a class of newly independent, urban women. Between 1900 and 1910, the number of women working outside the home increased by 50 percent. Many moved to cities, where one in five lived alone, supporting themselves on what were often abysmally low wages as shopgirls in department stores or line-workers in factories; politicians and journalists often referred to them as "women adrift."

Unlike their mothers and grandmothers, this new generation was not dependent on a father or husband for food and shelter. Enamored of their new social and economic freedoms, they demanded political power (the Nineteenth Amendment, enfranchising women, became part of the U.S. Constitution in 1920) and a modicum of sexual liberation, too—Stern writes that "rates of premarital sex rose sharply," and, by 1920, men reported having had double the number of sexual partners as their peers in 1910.

For many, these women posed an imminent danger—not just to America's unwitting and unfortunate men, but also to the innocent wives and mothers these men would subsequently infect, as well as their unborn children. The New York Board of Health claimed that 95 percent of so-called easy women carried an STD. No one seems to have seriously considered the idea that men could protect themselves by using a condom, patented in 1844, although the U.S. Navy helpfully removed doorknobs from its battleships, just in case syphilis could be transmitted by touch alone. Still more significant, though not explicitly stated, was the underlying fear that the very fabric of society was threatened by these women, who, Stern writes, seemed determined to follow "their own desires, dreams, and choices."

From Kansas court records, Stern put together a list revealing just how thin the grounds for quarantine could be. "One woman owed rent to a former sheriff, who had her taken in on suspicion when she could not afford to pay," he writes. "Another was arrested after changing jobs, when her former boss vengefully reported her to the health officer." Ms. A, a waitress, was detained on suspicion after dining alone in a restaurant. Women who were known to imbibe alcohol were frequent targets for the morals squad: detainee #3798 was brought in when police found whiskey in her possession, despite the fact that it had been prescribed by her doctor as a treatment for tuberculosis. Some women were even turned in by their husbands, as a way to exact revenge following an argument. A third of the women locked up under the Plan in Kansas were black—in a state where African Americans made up only 3 percent of the population.

Women were quarantined under the American Plan without bail or due process. Technically, the law's provisions could have applied to men, too, but, when asked whether it was being applied in a discriminatory fashion, Gardener M. Byington, a health official in Michigan, waved away concerns, explaining, "Our Department feels that a female can spread Venereal Disease a great deal more rapidly and, usually, it is easier to hospitalize a female than a male owing to the fact that the latter is a wage earner."

Unlike the case of the Chinatown plague, legal support was not forthcoming: when a woman named Billie Smith sued the government for locking her up on suspicion of having syphilis, the Arkansas Supreme Court ruled that "the private rights of [the] appellee, if any, must yield in the interest of the public security." Just as today, public jurisdiction over the female body was seen as perfectly acceptable, even by those who stood firmly against other forms of discrimination. As Stern notes, such liberal icons as Franklin Roosevelt; Earl Warren, who wrote the *Brown v. Board of Education* decision; and the American Civil Liberties Union all publicly supported the Plan.

⟷

Defined by suspicion, and thus inevitably subject to bias, quarantine has always been ripe for abuse. It has a dismal track record of discrimination to match. Could the use of quarantine be justified, given the lasting stigma it confers on the quarantined, as well as the mistrust it generates toward the medical professionals attempting to implement it?

"You're not going to find anybody who is pro-quarantine," Marty Cetron admitted. "But the truth is, our most urgent modern biologic threats have required us to roll back to our fourteenth-century tool kit." By the 1950s, quarantine had fallen so far out of fashion that the term disappeared from World Health Organization vocabulary; it has been revived and, with varying degrees of success, reinvented only in the past couple of decades. The story behind quarantine's return involves an AIDS activist turned public health official, a statistical analysis of the 1918 flu, and, ultimately, Kaci Hickox. But it begins with quarantine's premature retirement, in the years immediately following World War II.

Antibiotics, which became widely available only postwar, rendered many of the most dreaded diseases of the past quick and easy to treat. From bubonic plague to typhus, and tuberculosis to syphilis, the availability of effective treatments made the use of quarantine moot. Thanks also to toxic but powerful DDT-based vector control, and the introduction of rapid diagnostics as well as safe and effective vaccines to protect against illnesses such as measles and polio, by the 1970s, infectious diseases seemed like a solved problem. At the same time, starting with the civil rights movement of the 1950s and '60s, the United States experienced what legal scholars have termed "the rights revolution"—a series of landmark decisions that dramatically increased the American judicial system's protection of individual liberties.

By the 1970s, the Supreme Court had ruled in support of a series of new constitutional claims, establishing important precedents against discrimination on the basis of race, gender, and sexuality. Among those protections were strengthened due process rights, in both criminal and civil law.

The combination was powerful: the shift in concern away from contagious and toward chronic disease allowed health to be reframed in terms of individual lifestyle choices, while the countercultural zeitgeist prioritized personal freedom over social constraints. Quarantine, as a medieval instrument of sanitary state police power, seemed like a relic from another age.

Then, on June 5, 1981, the CDC issued the first official report of what was later named AIDS: five cases of a rare lung infection discovered in young, previously healthy gay men in Los Angeles. Less than a month later, this mysterious, new, and seemingly fatal disease was being called "the gay cancer." By 1982, the media had labeled the groups of people perceived to be an AIDS risk the "4-H club": homosexuals, heroin users, Haitians, and hemophiliacs.

Although it seemed clear that AIDS was spread only through the exchange of certain bodily fluids, scientists couldn't initially exclude saliva and mucus from that list. A poll conducted in 1985 revealed that nearly half of Americans believed it was possible to catch AIDS by sharing a drinking glass, and 28 percent thought the disease could be picked up from toilet seats, all of which led more than a third of the population to conclude that it was unsafe to "associate" with someone who had AIDS, even if there was no physical contact involved. In Indiana, a thirteen-year-old boy with AIDS was banned from school; in California, the state Realtors' association insisted that its members inform potential buyers if a house was owned by an AIDS patient; and, in one tragic case, firefighters refused to resuscitate a man they thought might be gay.

Fear of and discrimination against already-marginalized groups at the epicenter of the epidemic were thus widespread, and with them came calls for good old-fashioned quarantine. Within a decade, twenty-five states had put in place measures to detain individuals suspected of transmitting HIV—many of them based on American Plan laws.

The federal government joined in, instituting a travel ban that denied entry to the United States to noncitizens with HIV or AIDS. That ban remained in place until 2010 and led to the creation of the world's first and only "HIV prison camp" in 1991, set up by George H. W. Bush and Attorney General William Barr. (Barr, who also served as attorney general in the Trump administration, refused to quarantine himself following a potential exposure to the coronavirus in October 2020.) The camp was built to house 274 Haitian refugees who had tested positive for the virus on the military base at Guantánamo Bay—itself a legal gray zone, operated by the United States but not within its national borders. This extraterritorial status enabled what was otherwise an illegal public health detention; the base would later house another set of extraordinary detainees, the unlawful enemy combatants from George W. Bush's War on Terror.

Meanwhile, the Justice Department ruled that federal contractors were permitted to fire employees with AIDS based on fear of contagion, a decision that was contrary to CDC advice and that prompted widespread criticism.

As Charles Krauthammer, writing in *The Washington Post*, put it, "It should not matter if people think you can get AIDS in the Xerox room. You can't. Ignorance is a cause of discrimination. It is not a justification for it." But, as Krista Maglen pointed out to us, the AIDS epidemic provides a clear example of yet another of quarantine's unusual powers: it can manipulate the public understanding of a disease. "A disease might not be transmissible person-to-person, or it might not be highly contagious, but the imposition of quarantine automatically implies that if I was sick and I stood in the same room and breathed on you, you would get sick," she said.

In California, which voted down a closely contested ballot initiative to classify AIDS as a quarantinable disease, the state's chief of infectious diseases suggested posting signs on the homes of infected individuals instead. The conservative commentator William F. Buckley Jr. wrote a *New York Times* op-ed calling for all HIV-positive individuals to have their status tattooed on their forearms and buttocks, to warn off anyone who might think of sharing a needle or having anal sex with them.

The attorney Mark Barnes was studying law at Yale in 1983, and he remembers debating the issue in class. A sex worker in New Haven, Connecticut, had tested positive for AIDS but continued to ply her trade. Occasionally, Barnes told us, the police would pick her up for prostitution, or she would spend some time in the hospital, but, each time she was released or discharged, she ended up back on the street. The class was taught by Virginia Roddy, legal counsel for Yale New Haven Hospital, and Angela Holder, whom Barnes described as one of the greatest research ethicists of her generation. "The question was whether there would be a reason to quarantine this woman," Barnes said.

"I remember that Virginia and Angela were very upset about it because they also didn't know what in the world to do."

Barnes, who, according to family lore, is a direct descendant of Daniel Boone, went on to found the AIDS Law Clinic at Columbia University in order to represent patients with AIDS in antidiscrimination cases. He became one of the leading AIDS advocates in New York City—"and one of the last ones who survived, actually, because I was one of the few that didn't have HIV," he told us. "Most of the people I knew at the time are dead."

Barnes became a New York City health official in the early 1990s, joining a department with the power to mandate quarantine and isolation. He and his colleagues quickly found themselves in need of these dangerous measures thanks to the spread of a new, harder-to-treat strain of an old contagious disease: multidrug-resistant tuberculosis (MDR-TB). Resistance occurs when patients are prescribed antibiotics, take them for a few weeks until the tuberculosis bacteria in their body are suppressed—so they feel better—and then stop before actually wiping out the infection. Their symptoms often return a few months later, the bacteria having evolved defenses against that antibiotic. Eventually, thanks to thousands of people failing to take the full course of antibiotics as prescribed, some strains of tuberculosis have ended up impervious to all the available antibiotics, and a curable infectious disease has become incurable once again.

In New York City, between 1978 and 1992, there was a nearly 300 percent rise in tuberculosis cases. (TB is one of the most serious infections in people with HIV/AIDS.) Nearly a third of them were drug resistant. "Most public health experts think that when you get to that point in a population, at least in tuberculosis, you end up with a situation in which the entire system could spin out of control," said Barnes. Up to eighty percent of people with extensively drug-resistant tuberculosis (XDR-TB) will die, and they can spread the infection to others by coughing and sneezing.

One proven preventative strategy, known as directly observed therapy, is to have public health workers watch each patient take their medication, every single day, to make sure they complete the course and don't develop resistance. New York City used this technique, and the department developed a range of incentives—meals, transportation expenses, referrals to other services—to increase retention. Still, at least 10 percent of the city's new TB diagnoses were "noncompliant"—often individuals with unstable lives, whether due to addiction, mental illness, or lack of shelter, that made it difficult to complete therapy.

"So the question was: What do you do with those people?" Barnes said. New York had quarantine authority written into its health code, of course, but that section hadn't been updated since the rights revolution, and failed to include either a standard for determining whether detention was appropriate, or procedural protections for the detainee. To Barnes, it seemed unusable. He turned instead to the state's statutes for the civil commitment of the mentally ill. These had been heavily litigated during the rights revolution and now included a web of constraints on the state's power, such as the requirement that health officials base their decision on the best interests of the individual as well as the public, and guaranteed rights for the detained individual, including the provision of free counsel and judicial review.

Over the next six months, Barnes worked with colleagues to draft a new quarantine and isolation statute with the mental health model in mind. "Once a week for about an hour, I invited the most vociferous civil liberties people into my office at the health department," he told us. "I sat with them and went through the draft every week, soliciting their opinions, telling them what was possible and not possible, and listening to them." The end result dictated that the state must do everything it reasonably could to achieve voluntary compliance. Still, compulsory detention was a permissible option. The health department set up a secure unit with twenty-five beds in a wing of the now-demolished Goldwater Memorial Hospital on Roosevelt Island, where patients could be "sequestered," as Barnes put it, while receiving three meals a day, housing, mental health and rehabilitation services, and, of course, their TB regimen.

"All the naysayers said, you can't lock up three thousand people—you don't know what you're doing," remembered Barnes. "And I said, no, but if there is a credible quarantine civil-detention alternative, it will reorder the rest of the world, and they will start taking their medicine. And that's exactly what happened." Within two years, he told us, New York City had broken the back of its MDR-TB epidemic.

⟷

In the early 1990s, Mark Barnes's efforts to reform quarantine were the exception rather than the rule. However, in the post-9/11 panic, fueled by the subsequent postal anthrax attacks, and the 2003 SARS outbreak, as well as by the Bush administration's false allegations that Saddam Hussein had assembled drones full of bioweapons, the need to update state and federal quarantine powers took on new urgency. The nation looked at its public health detention statutes and realized, as Mark Barnes and Marty Cetron already had, that they were deeply flawed: outdated, inconsistent, and dangerously overbroad, as well as occasionally completely irrelevant, as in the case of a New Jersey law forbidding common carriers from transporting infected bedding.

Cetron and his colleagues had begun tinkering with some draft legislation in the late nineties, with the idea of creating a template for individual states to adopt. In September 2001, at the CDC's behest, the public health law professor Larry Gostin reworked the draft in four weeks flat, producing the Model State Emergency Health Powers Act—much of which, Barnes noted, was heavily influenced by New York City's example.

Over the next few years, thirty-three states began the legislative process of incorporating some or all of its provisions into their health code. In Atlanta, meanwhile, Cetron immersed himself into quarantine research, determined to transform this newly in vogue medieval tool into a modern public health practice. The first question he needed to answer was: Does quarantine actually work? And, if so, under what circumstances?

Cetron partnered with the author Howard Markel to apply cutting-edge statistical analysis to the 1918 Spanish flu pandemic, reconstructing the impact of various interventions—isolating the sick, quarantining the exposed, closing schools, canceling mass gatherings, and so on—on the spread of the virus in forty-three different cities, at a range of different times along the epidemiological curve.

"Most people don't appreciate that quarantine principles and practices are not one thing," Cetron told us. "It's a full spectrum, and that spectrum goes in many directions." For Cetron, school closures or mask ordinances lie on the same axis as quarantine and isolation—they are all just more or less restrictive social distancing measures. Similarly, these interventions can run from voluntary to compulsory, with varying degrees of incentives and enforcement in between, or from individually targeted to mass, population-level restrictions. "All of these are dimensions of the same tool," he explained.

Cetron and Markel discovered that the timing, the duration, and the layering of these various measures made a huge difference—not just to reducing the overall number of deaths, but to mitigating the pace and peak of the flu's spread. The cities that acted early and aggressively, closing schools and banning public gatherings at the same time, and maintaining those restrictions for the longest, did best—to use an expression that became mainstream during COVID-19, they "flattened the curve." Markel credits Cetron with coining the term, over "bad Thai food," back in 2007; although Cetron confirmed this, he characteristically used Markel's anecdote as an opportunity to highlight another essential public health principle: "This is a 'we not me' moment," he said.

Cetron uses the Swiss cheese model of risk analysis to explain how the various interventions on his quarantine spectrum work together: individually, each is as riddled with holes as a slice of Emmental, but if you stack enough of them, a virus will struggle to get through. That said, his analyses have also disproved the myth that quarantine needs to be complete to be effective. "Depending on the epidemic, you may be able to get fifty percent leakage and still quench it," he told us, explaining that a voluntary quarantine may contain a virus more quickly, despite its inherent leakiness, than more stringent measures that can drive cases underground.

The evidence from historical analysis and modeling studies seemed strong enough to justify the use of quarantine under certain circumstances. Cetron's next question was whether compulsory quarantine, a fourteenth-century practice in which individual freedoms are sacrificed for the benefit of others, could ever be compatible with twenty-first-century ideas about human rights and health equity. He partnered with Julius Landwirth at Yale to review the existing guidance on situations in which a person's liberty and freedom of movement can ethically be restricted, and then develop a framework for what they called "modern quarantine."

Cetron summed up their conclusions in what has become his quarantine mantra: three primary questions that he repeats at every opportunity. *May I*—do I have the legal authority to take this step? *Can I*—do I have the resources to implement and maintain these measures? And *should I*—am I sure this is the least restrictive way to achieve my public health goals, in terms of saving lives or slowing the spread of disease? "If you come up with a yes—you may, you can, and you should—then the most important aspect of it is the *how*," Cetron continued. He and Landwirth determined that any public health restrictions and enforcement should be proportional to the consequences, and equitably and transparently applied, without bias and with respect for individual privacy. Additionally, as in Barnes's New York MDR-TB statute, the quarantined must be guaranteed the right and the means to challenge their detention in court.

Most important, Cetron told us, anyone being asked to temporarily give up their own rights in order to protect the public is owed a duty of care from that public, to be provided by the taxpayer-funded CDC. "That means basic human needs—food and water and shelter," said Cetron, but it ought to also include more substantive support: a means of communication, medical and mental health treatment at no cost, and early access to diagnostics, therapy, and a vaccine for the disease in question. "There's no control without care," Cetron explained. "It's not health security versus human rights—it's a carefully negotiated balance between them."

Cetron incorporated his findings on the timing, duration, and layering of social distancing measures into the

CDC's strategy guidelines for mitigating pandemic influenza, issued in 2007 and updated in 2017. He made his reinvented "modern quarantine" into CDC policy in 2005, and pushed to amend the code of federal regulations, so that its new provisions and safeguards were spelled out in black and white—a bureaucratic process that took until January 2017. When we spoke to him after the new language became law, he seemed relieved. "I would sleep less comfortably if there was no other check than my own individual conscience and I had unchecked powers in this area," he said. "It's not a good position to be in."

In the United States, however, most quarantine powers reside at the state level, where governors sometimes overlook the last question in Cetron's quarantine catechism—*should I?* "The mistakes happen when people start with, 'I have the authority, so I may, and I have the capacity, so I can, therefore I should,' because there's all this pressure to do something," said Cetron. "And you get Chris Christie and Kaci Hickox facing off in a tent in New Jersey."

⟷

"I was driving to work that morning, fat, dumb, and happy," said Charlie LaVerdiere, retired chief justice of the Maine District Court. LaVerdiere was speaking at a national conference of state-court chief justices, gathered in Omaha, Nebraska, in May 2019, specifically to discuss how best to update and modernize each of their individual states' powers of quarantine. As he began to speak, the knot of court officials clustered around the coffee urns quickly returned to their seats: LaVerdiere had our full attention. "I'm just going to do my job, when I got a phone call that said, 'We have a problem,'" he continued.

The problem was Kaci Hickox. She had ended up spending three nights confined to her isolation tent in New Jersey. "I remember it being cold," Hickox told us, recalling the experience several years later. The tent was equipped with only a PortaPotty, she said, and there was no shower, but that was not why Hickox was unhappy. "I mean, when you're working in the field with Doctors Without Borders, you can sleep on the floor," she said.

Hickox's main complaint was that there was no real reason for her to be in quarantine. As terrifying as Ebola is—"it is a *scary* disease," confirmed Hickox—it is not contagious until you are actually sick. Hickox had no symptoms—she had an elevated temperature according to an infrared thermometer, but a normal one using the much more accurate oral measurement—and her blood test had come back negative.

"Now, the news media portrayed her as a nurse that had gone to West Africa and was exposed to Ebola and brought it back to the United States," said LaVerdiere, a solid, pragmatic individual with a trim white goatee and exquisitely deadpan delivery. "Some of which was true." Hickox was indeed a nurse, but she was also, like Marty Cetron, a graduate of the CDC's elite Epidemic Intelligence Service program and had worked with MDR-TB patients in Las Vegas. Hickox understood Ebola transmission, and she knew how quarantine was supposed to work. "It's often homeless people who end up having multidrug-resistant TB, and we'd help them find a place to live and do direct observation therapy for their treatment," she told us. "A quarantine conversation would be the very last resort."

Her own expertise meant that, like many a well-meaning individual before and since, Hickox had overestimated the public's ability to make rational decisions based on scientific evidence, and underestimated the fear and hysteria inspired by epidemic disease. "I will say that I was pretty naive," she admitted. Her mother had actually texted her while she was in Brussels, debriefing with the Doctors Without Borders team, to ask whether she thought there might be an issue when she flew in. "LOL that's not going to happen," was Hickox's response.

In fact, when Hickox spoke on a network news show on Sunday evening, her third night in the isolation tent, she alienated many with her calls for evidence-based decision-making. "Ms. Hickox demanded that an undereducated America worship at the altar of the 'science' that she invokes to argue for her immediate release," sniffed the Heritage Foundation–funded website Townhall.com. Americans "deserved better" than the "snotty" and "haughty" comments of a woman who dared to know what she was talking about and defy mainstream opinion.

By Monday, Hickox's lawyer, Norman Siegel, had persuaded Chris Christie to release her—as long as she left the state. Hickox would no longer be New Jersey's problem. Yet even her means of transport was subject to petty wrangling: Siegel volunteered to pick her up, but officials insisted she make the ten-hour journey in an ambulance, which Hickox resisted. "As soon as you get into an ambulance, it looks like you're ill," she pointed out. "And wouldn't that be just what Governor Christie wanted for the public to think I was?"

Eventually, Hickox was driven home to Fort Kent, Maine, in an unmarked black SUV. She had moved to the tiny town on the Canadian border that summer, just a month before she left for Sierra Leone; her partner, Ted, had enrolled in the nursing program at the University of Maine campus there. "I was in between jobs and Doctors Without Borders sent an email and said, we really need people with outbreak experience, please come," said Hickox. "Ted and I were on a canoeing trip and I said, 'I feel like I should go,' and he said, 'Yeah, you absolutely should go.'"

While Hickox was en route from New Jersey, Maine's governor, Paul LePage, who was also in a tight reelection fight, decided that she should serve out the rest of Ebola's twenty-one-day incubation period under quarantine at home. For the first couple of days, Governor LePage sent state troopers to monitor her house and health workers to watch her take her temperature every day.

Then, on Thursday morning, Hickox and Ted left the house for a bike ride.

"I think so many people thought, oh, that nurse, she's being arrogant and flamboyant by going out on a bike ride," said Hickox. "But truthfully, leaving the house was the only way I could force them to go through their own process so that I could have due process." Through her lawyer, Hickox had requested that, if the state of Maine was going to require her to stay at home, they should issue a formal quarantine order, so that she could challenge it in court. But it wasn't until she left the house that Charlie LaVerdiere got an urgent call.

"All of a sudden, on a nice October morning, with the leaves all turning, a ton of stuff from the skies falls all over me," LaVerdiere told us. He was stuck: if he didn't issue a ruling, Hickox had said she was going to head out into town for pizza. "No decision was, in fact, a decision," LaVerdiere realized, which meant that he needed to set up an interim hearing right away. Logistically, this posed a series of hurdles. Fort Kent is a five-hour drive from the state capital, in Augusta, and its tiny courthouse was staffed with a part-time judge and two clerks.

"We began to understand, wow, we've got a lot of issues here in terms of how we're going to do this," said LaVerdiere. "Because we've got to record all the hearings. We've got to give her an opportunity to testify. She doesn't have internet at home. Boy, this is going to get complicated." Meanwhile, those two clerks had begun receiving thousands of calls from reporters around the world. "I can tell you that the media was in a frenzy," LaVerdiere said, "and we suddenly have clerks and court marshals that were saying, not me, I'm not going to go to that court for that hearing—I've got family."

A larger challenge was the legal framework itself—or lack thereof. LaVerdiere and his staff looked for precedents and substantive law to inform their process. "It took us a fair amount of research to find out that the Maine statutes were absolutely confusing and completely contradictory, and the national cases we looked at were all old and not on point," said LaVerdiere. "We even had questions as to who had the burden of proof on issues."

The clock was ticking, so LaVerdiere played for time: he roped together an interim hearing over the telephone with the various attorneys involved, then issued a temporary status quo order so that Hickox could not leave her house for twenty-four hours. Then, he said, he tried to get his head around both the law and the science. "I didn't get a lot of sleep that night," he admitted.

The next morning, LaVerdiere gathered the attorneys again and told them that, in his opinion, the state had not met its burden, and thus he could not allow its request for quarantine. In his ruling, LaVerdiere decided to add a couple of paragraphs of commentary, noting that Hickox was owed a debt of gratitude for her kindness and compassion in caring for Ebola patients, and that the court was fully aware that the hysteria motivating her initial quarantine was not necessarily justified. "However, whether the fear is rational or not, it is present and it is real," LaVerdiere wrote. "Respondent's actions at this point as a health care professional need to demonstrate her full understanding of human nature. She should guide herself accordingly."

"I, uh, may have gone a little further than I should have," LaVerdiere said, after reading his commentary to us. "But I did this in the hopes that Kaci Hickox would get two messages: yes, you won, and you are free to leave—but if you're smart, you'll stay home." LaVerdiere paused for a moment, before delivering the denouement. "She stayed home."

"I thought that his ruling was very eloquent and thoughtful," said Hickox. "And I'm just thankful because I also imagine that he was getting a lot of pressure. I think for him to make the decision to do what he believed was right under the law—that takes courage."

To this day, Hickox and LaVerdiere have never met, or even spoken. "The conversations were completely between my lawyers and the Maine state health department and him—I never heard his voice," said Hickox. "I've never been able to actually thank him."

Hickox continued healthy and symptom-free, but the entire episode was not without lasting costs for everyone involved. "For months after I issued the decision, I was still getting death threats," said LaVerdiere. "We also had lots of political fallout." Governor LePage, who had made protecting Mainers from Kaci Hickox part of his campaign for reelection, was particularly unhappy. "We don't know what we don't know about Ebola," he said. "But the monkey's on [the judge's] back, not mine."

Hickox also received her share of hate mail. "I got this one letter that was four times the size of a normal piece of paper, and it said, 'You arrogant cunt bitch, you give nursing a bad name,'" she remembered. "And another letter that said, 'I hope you get Ebola and die.'" Locally, she said, many people were supportive—strangers dropped off groceries at their door, and a couple of Ted's nursing classmates brought over treats. "I'd be like, I just need a beer," Hickox said. "So they'd bring me a six-pack."

Nonetheless, she and Ted ended up moving out of state shortly after. "By that time, that house had so many negative connotations for us," said Hickox, and, in any case, Ted had dropped out of his nursing program, frustrated by what he saw as the school giving into unscientific hysteria by insisting that he stay away from campus. "I wish I could have communicated more a sense of calmness," Ray Phinney, associate dean of student life and development at the university, told Maine Public Radio a year later.

Beyond the impact of Kaci Hickox's quarantine on individual lives, the policy had what Doctors Without Borders called a "chilling effect" on its ability to recruit desperately needed volunteers to help care for Ebola patients in West Africa. A monthlong volunteer assignment would be nearly doubled in length by the twenty-one-day quarantine rules, and doctors and nurses were unwilling to inflict the associated stigma on their families.

Hickox's original hope was that by challenging her quarantine she could help avoid this outcome. "I have the kind of personality where I will assert myself when I need to," she explained to us. "And I was thinking of some of my colleagues who maybe are not as assertive, and thinking, if another healthcare worker comes and has to go through this, then that is going to be devastating."

In the long run, however, Hickox may have achieved her goals. A year later, with the support of the ACLU of New Jersey, she sued Chris Christie for false imprisonment. The case was eventually settled—but rather than fight for monetary compensation, Hickox made her agreement conditional upon the implementation of a new "bill of rights" for the quarantined in New Jersey: a detailed set of procedures spelling out everything from how temperature readings should be taken to the rights of the quarantined to challenge their detention in court. The resulting document closely resembles Marty Cetron's own updated federal quarantine statute, but with even more specificity about the steps public health officials should follow to arrive at a medically appropriate decision, to inform individuals of their rights, and to demonstrate that less restrictive means have failed or are not appropriate.

"I'm a public health nurse, so I know that sometimes quarantine will be needed," said Hickox. "But when we do it, we need to do it well and we need to think about that person as a human with a family and a livelihood and everything else." If those principles aren't in writing, she told us, politicians will abuse their powers of quarantine and isolation, whether out of ignorance, malevolence, or simply in order to seem tough on disease. "It's really easy to trust our gut instinct, which says just stay at home for three weeks and stop complaining," Hickox said. "But the negative unintended consequences of bad policies are large."

5

Alone Together

The first federal quarantine facility constructed in the United States for more than a century is a retrofitted parking garage in Omaha, Nebraska. "Much to everyone's chagrin," said Ted Cieslak, who serves as medical director of the National Quarantine Unit, "because parking is hard to come by on campus." When we visited on a rainy, gray day in May 2019, major construction was over but UV-reflective paint—a soothing palette of light gray and teal—was still drying on the walls. Hard hats were mandatory as electricians, plumbers, and HVAC engineers took care of the final details. Each of the twenty rooms had a desk, closet, and en suite bathroom with a recessed vanity and shower: we could have been in an Econo Lodge, but for the vinyl flooring that extended two inches up the wall (for more effective disinfection) and the individual control boxes outside each room whose screens read "Negative Pressure Disabled."

"It's more like a hotel than a hospital," agreed Cieslak, a blunt, good-humored epidemiologist with a ready laugh. Cieslak came to the University of Nebraska Medical Center after thirty years in the military, at the U.S. Army Medical Research Institute of Infectious Diseases at Fort Detrick, Maryland. There, he was responsible for one of the nation's

A hallway inside the then-under-construction National Quarantine Unit in Omaha, Nebraska. The white control boxes on the walls regulate the internal air pressure of each individual quarantine room. (*Photograph by Geoff Manaugh*)

first biocontainment units—"the Slammer," so named for the ominous sound its steel doors made while closing. "We never had to use that as an isolation facility, fortunately," said Cieslak, "but, over the years, we did admit twenty-one people for quarantine." These unfortunate few had somehow been exposed to a pathogen—for example, when an Ebola-infected monkey managed to rip a hole in their protective suit—but were not yet known to be infected.

"You can imagine the psychological angst," Cieslak said. "These people have just had a very frightening exposure, and they're immediately taken and put into quarantine in the Slammer, not knowing if they will ever see their family again, because, if they do get sick, they're going to stay there." This, he explained, was the quarantine paradigm for such high-level pathogens in the past: an experience made notorious by Richard Preston in *The Hot Zone*. People in the Slammer, Preston writes, "start flaking out by the second week. They become clinically depressed." Some, Preston continues, "become agitated and fearful," some so angry and paranoid that they "need to have a continual drip of Valium in the arm to keep them from pounding on the walls."

"Eventually that tends to give way to boredom," Cieslak told us. "It's incredibly boring to be in quarantine, because you feel well, but there's no way to get out and get any fresh air, have a beer, or do whatever you want to do."

Sadly, in the new National Quarantine Unit, beer will still not be an option: the campus is dry. Apart from that, efforts have been made to "ameliorate the situation," as Cieslak put it. Rooms are equipped with Wi-Fi, flat-screen TVs, minifridges, even exercise bikes. The windows, however, do not open, while the doors have no locks: if a person under quarantine develops symptoms, staff wearing personal protective equipment must be able to enter and move them to the university hospital, and its high-level biocontainment unit, conveniently located across the street.

Before the National Quarantine Unit opened for business in January 2020, the CDC operated quarantine posts at airports and land-border crossings in order to screen incoming travelers for infectious disease risk. It also possessed the legal authority to order those individuals into quarantine. What it did not have was anywhere to put them. From the 1950s, as the volume of international travelers arriving by sea plummeted due to the rise in air travel, and the threat of infectious diseases seemed to recede, maintaining America's increasingly decrepit lazarettos seemed like an unnecessary expense. The U.S. government retained its quarantine powers, but their exercise was largely left to state and local authorities.

Some former federal quarantine stations, such as at Ellis Island in New York Harbor and Angel Island in San Francisco Bay, have since become popular tourist sites—places to experience the anticipation and optimism of immigrant ancestors briefly delayed on the brink of arrival in the promised land. Others have not been as celebrated: the quarantine station in Baltimore is now home to the Quarantine Road Sanitary Landfill, a Superfund site; and all four structures used to quarantine enslaved Africans on Sullivan's Island, in Charleston, South Carolina, have long since disappeared. The oldest surviving lazaretto in the Western Hemisphere is in Philadelphia. As we toured its crumbling but still elegant colonnaded facade with David Barnes, the historian leading the fight to preserve it, he explained that, following the quarantine station's closure in 1895, it had been repurposed as a post office, a retirement home, a training ground for the Philadelphia Athletics baseball team (now the Oakland A's), and an aviation school.

Around the world, the same story is true: the "spacious" and "commodious" lazaretto at Marseille that so impressed John Howard was demolished without fanfare to make way for a new commercial port with deeper docks in the 1850s; the magnificent lazaretto in Malta became an ordinary hospital in 1949, closed its doors in the 1970s, and descended into its current ruinous state while serving as a dog shelter during the subsequent decades. Even in Australia, where enthusiasm for biosecurity was sustained, the sprawling quarantine station at Sydney's North Head finally ceased operations in 1984, only to reopen as a spa-hotel in 2006, its remote location making it the perfect retreat from the city.

Starting in the 1980s, healthcare in the United States moved away from housing patients during treatment, as part of a larger shift known as deinstitutionalization. The national leprosarium in Carville, Louisiana, where individuals suffering from Hansen's disease had been held for life, closed its doors in 1999; the dozens of sanatoriums for tuberculosis patients, built in states considered to offer a salubrious climate, were repurposed or shut down. Advances in medicine meant that these chronic contagious diseases were now curable, and outpatient care was a much more cost-effective option for hospital administrators focused on maximizing profits. At the start of the new millennium, even the Slammer's future was on the line. It was entering its fourth decade but had never been used for its original purpose—isolation. It had also become both expensive and difficult to maintain. "The guy who made much of our equipment—he was a one-man shop and he died," said Cieslak. For a while, it seemed possible that the United States might have nowhere to isolate people who either were, or could be, infectious.

Then, as a series of scares over the next few years—9/11, the anthrax attacks, and SARS—refocused the nation's attention on its vulnerability to pathogens, the disinvestment in public health infrastructure over the preceding decades began to look like a mistake.

"After those anthrax attacks, Congress did what Congress does," Cieslak told us. "They appropriated a lot of money." Most of the fifty states spent their share of the biodefense windfall to purchase drugs and vaccines; Nebraska decided to build a high-level biocontainment unit. Cieslak's predecessor, Dr. Phil Smith, realized that what the United States was missing was somewhere to isolate and treat patients with contagious, highly pathogenic diseases, and managed to convince the state legislature and local emergency services that Omaha was the place. "There were a lot of places at the time that had the attitude that, not in my backyard, you're not going to bring Ebola," said Cieslak. "But the Omaha Fire Department embraced this, and now it's a badge of honor—they are the experts in the transport of highly hazardous communicable disease patients."

When the first Ebola patients arrived in Nebraska in 2014, Smith's foresight was validated. But, given the number of American healthcare workers who, like Kaci Hickox, had volunteered in Sierra Leone, Liberia, and the Democratic Republic of the Congo, and the furor surrounding their return, the Department of Health and Human Services decided that the nation's capabilities were still not robust enough. They dished out some more money, including a $19.8 million grant to the University of Nebraska Medical Center in November 2016, to build the nation's only federal quarantine facility, smack in the middle of the country.

"We're roughly equidistant for both coasts," Cieslak pointed out. "In his proposal, Phil Smith sold that as a logical reason to locate it in Omaha." Once confined to borders and edges—ideally, at a safe distance offshore—the twenty-first-century lazaretto has been relocated to the heartland. Rather than use distance to neutralize the threat of an escaped pathogen, the National Quarantine Unit relies on HVAC, personal protective equipment, and digital keycards.

As our tour of the new building continued, ascending a level to the visualization and training facility, complete with a holographic Ted Cieslak, we asked the real version whether twenty beds would be enough for the entire nation's quarantine needs. "We've gotten by forever without any federal quarantine facility, so I'd hesitate to say that we need more," he replied, as we spoke in 2019. "My gut feeling is, yes, it's enough." There just aren't very many deadly diseases that are contagious, incurable, *and* potentially transmissible before they cause symptoms, in order to justify quarantine, he explained.

Based on his experience up to that point—managing one or two people at a time with potential exposure to something like Ebola—the new quarantine unit was ideal. "If it were to go beyond that—if we were to have hundreds of people that need this—then I think the horse is out of the barn," Cieslak said. "Now we're talking about a pandemic."

⟷

On January 31, 2020, just days after the National Quarantine Unit had been certified as operational, Marty Cetron at the CDC signed the first federal quarantine orders in decades, authorizing the mandatory detention and observation of hundreds of U.S. citizens repatriated on State Department flights from Wuhan, China. A couple of weeks later, they were joined by another four hundred Americans evacuated from their coronavirus cruise aboard a ship called the *Diamond Princess.*

Cetron heard about the new virus in China over the New Year's holidays, while vacationing with family in New Hampshire. Over the following weeks, even as George Gao, the director of the Chinese CDC, assured his American counterparts that, based on reports from Wuhan health officials, human-to-human transmission was not occurring, Cetron's concern began to build. He also recognized a fatal tendency from his study of outbreaks past: a coping strategy based on wishful thinking. "If you recognize how big it is, and, if you really recognize asymptomatic transmission is a driver, then you have a problem on your hands that is absolutely overwhelming," Cetron told us a few months into the pandemic. "In general, people would rather box it up—keep the focus on something tight and small."

Still, the decision to order a compulsory quarantine gave him pause. "It was the largest population-based quarantine under U.S. federal authority in over one hundred years," Cetron said—but, he reasoned, it made sense as a way to buy some time. The emergency repatriation flights had been organized almost overnight, leaving no time for public health workers to conduct assessments of each individual's risk of exposure or symptom history. As part of the process of modernizing federal regulations, Cetron had deliberately built in a safety valve: all quarantine decisions were subject to a mandatory review after seventy-two hours. "The original plan was to repatriate out of Wuhan, take seventy-two hours to assess people in the U.S., and then we'll see what we're really dealing with," Cetron explained. "Maybe a lot of them could have gone home; maybe there were risk-mitigation options short of a wholesale, military-style quarantine."

By this point, Cetron and many of his colleagues were relatively certain that a pandemic was inevitable. "There wasn't some hope that we were going to wall this off in China or any other country," he told us. The first U.S.-based case of COVID-19 had been diagnosed in Seattle on January 20, and by late February, community transmission was confirmed to have taken place. Nonetheless, amid the uncertainty of those early days, even Cetron found it difficult to make the mental leap required to give up on trying to use entry-screening and quarantine to prevent the virus from gaining a foothold in the United States, and, instead, focus on reducing viral spread in the community, through strategies such as school closures, mask mandates, and lockdowns. Political leaders were, understandably, more reluctant to make such a shift.

To Cetron's evident regret, under political pressure, his more measured quarantine plan did not even survive the full seventy-two hours. "All of a sudden," he told us, "we were knee-deep in a massive federal containment effort." In early February, several thousand seemingly healthy people were being detained under Cetron's signature, and he had ceased to be able to sleep at night. "How do we live up to our own regulatory standards of preserving equity and using the least restrictive means to achieve the public health goal in this situation?" he asked. "These are enormous, enormous dilemmas."

The first flight out of Wuhan carried 195 people, making the National Quarantine Unit's brand-new twenty beds seem wildly insufficient. That flight was quickly followed by several more. The hundreds of men, women, and children eventually evacuated from China had to be housed at military reserve bases around the country for the duration of their quarantine. But, when the fifteen Americans aboard the *Diamond Princess* who had tested positive for COVID-19 needed somewhere to go, the National Quarantine Unit received its first occupants. This wasn't quarantine, strictly speaking—the passengers were known to be infected—but the unit seemed like the best place to isolate them until proven safe.

When we spoke to Ted Cieslak and his colleague Rachel Lookadoo, who is responsible for the University of Nebraska Medical Center's legal and public health preparedness, they told us that the facility's debut had, all things considered, gone well. "The quarantined individuals were fantastic—they were very cooperative, very agreeable," said Cieslak. "I would have gone nuts in a day, but they all accepted it pretty well." One couple were newlyweds; Cieslak told us that they were allowed to room together in what was henceforth known as the quarantine unit's honeymoon suite. Apart from that, the group's only interaction with one another was in the form of a one-hour daily "town hall" call, run by the unit's behavioral health consultant, the psychologist David Cates. "Every day he would give them a new coping strategy and talk to them about that," said Cieslak. "And they had access to Amazon, so they could pretty much order anything they wanted."

One issue that the team at Nebraska hadn't thought through when designing the quarantine facility was that, because it is on the ground level, members of the media were able to evade security, walk up to the building, and peer in through the windows. "At first, we were worried that it would be a challenge to the individuals' privacy," Cieslak said. As it turned out, at least two of the unit's guests, a married couple who owned a radio station in Santa Clarita, California, welcomed the attention—to the extent that, in the end, it was the unit's staff whose privacy was compromised. "We had a little mini media event, with the media parked around this one guy's window, him broadcasting through the window via his iPad to his own radio station," recalled Cieslak, laughing ruefully. "I still have their station bookmarked—I listen to it sometimes when I'm getting ready for work."

Overall, the couple—Jeri Seratti-Goldman and her husband, Carl—gave the unit broadly positive coverage. In a video broadcast from her quarantine room, Jeri bemoaned the lack of snacks, fresh air, and human contact, reporting that "they would come into your room only three times a day for food and temperature checks, and it's lonely." Nonetheless, she told viewers: "I took it as a gift. I used this time to reflect on my life. I even started doing adult coloring books, which was a blast."

In order to be released, the quarantined individuals had to test negative for coronavirus on three consecutive occasions. For some, this took weeks. Meanwhile, as February became March, much of the rest of the country began to shelter in place, with coronavirus infection rates in hard-hit New York City rapidly approaching those in Wuhan itself. "I always kind of felt like I had a tiger by the ears," said Cieslak. "I'm sure every one of these individuals would have much preferred to have quarantined at home, but, once we had them, we felt like we couldn't let them go." Although their fellow citizens were now isolating at home en masse, Cieslak couldn't legally release potentially infectious individuals from the unit and drop them off at a rental car agency. "There was a little bit of grumbling," said Cieslak. "But we all were trying to build the airplane while we were flying it."

As Lookadoo pointed out, the CDC was implementing regulations that had been recently updated, thanks to Cetron, but had never been tested. Cieslak had no idea what to do if his charges refused to undergo yet another nasal swab; Lookadoo was similarly uncertain as to who was actually going to pay for their expensive, involuntary medical detention. "The regulations state that the CDC *may* pay for things, but not necessarily that they *will* pay for things," she pointed out. "I think that was one of the many areas where it became clear the answers weren't well established beforehand."

Perhaps most obviously, but importantly, the kind of quarantine envisioned by the CDC's updated regulations and the brand-new Nebraska facility is a small-scale one. Both were clearly designed with the last pandemic—Ebola—in mind. The federal quarantine experience might have been short on snacks and fresh air, but in return for depriving a few cruise-loving Americans of their freedom of movement, the government supplied food, medical care, compensation, and job protection rights. Quarantine worked—for fifteen people.

Delivering on those promises for all 245 million Americans living under some form of lockdown by the end of March was a much more expensive and logistically challenging proposition—which is undoubtedly why those restrictions were not presented as quarantine orders, even though they were equally designed to deprive individuals of their freedom of movement in the interest of the common good. The contrast between the boutique experience of federal quarantine, managed by a trained staff of dozens, and the ad hoc support provided by food banks, community groups, and neighbors to hundreds of thousands of Americans who couldn't afford to shelter in place, was jarring.

"Especially when we look at the people who are being affected by COVID the most, it's frequently people who are the most vulnerable," said Lookadoo. "To have those rights ensured is important." What's more, added Cieslak, "sending a person home to quarantine when they've got other folks living in the house who are potentially susceptible is, I think, a big problem."

According to researchers who followed families in Nashville, Tennessee, and Marshfield, Wisconsin, if one member of a household catches COVID-19 and isolates at home, on average more than half of the people they live with will get infected. For communities in which multigenerational families live together in small houses and apartments, the consequences have been devastating: a daughter who ends up burying both parents and her own husband while caring for her children and recovering from the virus herself; a man whose father and two sons all die from COVID-19 while he is in the hospital struggling to breathe. In countries with lockdown measures in place, household transmission was estimated to be responsible for 70 percent of all new COVID-19 cases—the virus's public control achieved at the cost of domestic tragedy.

"I don't know an easy answer," said Cieslak. "Clearly, we can't build massive quarantine facilities to deal with all that." He paused, struck by the idea. "You *could*, I suppose."

⟷

When America's leaders need something built on the fly during an emergency, they call the Army Corps of Engineers. In March 2020, when Andrew Cuomo, the governor of New York, realized that the state was experiencing such an exponential surge in COVID-19 cases that it was on track to run out of hospital beds in a few weeks, he asked Lieutenant General Todd Semonite, the fifty-fourth chief of engineers, for help.

On the plane ride to Albany from Washington, D.C., Semonite and his staff came up with a five-page concept: a standardized design to convert existing building typologies found across the United States into something that was not quite a hospital, but close enough. These "alternative care facilities" would come in two main flavors, which Semonite referred to as "small room" and "big room."

"Small room," he told us, "would be college dormitories or hotels like you find on the side of a highway—think of a Fairfield Inn." In this instance, the mind-numbing, déjà vu–inducing sameness of the American built environment revealed an unexpected benefit, allowing the army corps to develop an eight-ingredient foolproof formula for flipping these cookie-cutter hotels. "A real simple example is taking out carpet," said Semonite. "The medical guys hate carpet because it's very, very hard to clean and sanitize." A slightly more complex retrofit involved hacking the hotel's HVAC system to create a negative-pressure environment for infection control. "We actually went into one of the Marriott hotel rooms and looked at the system," Semonite told us. "Turns out, you can actually trick the air conditioner to make that pressure so low that it would be able to keep the virus in."

"Large room" conversions—the transformation of a convention center, indoor athletics stadium, basketball court, and even a dressage arena—involved a little more creativity. Indeed, in the case of the horse-riding facility, it involved removing more than thirty-three thousand cubic feet of dirt. "I don't know much about horses," Semonite said. "Believe it or not, the dirt is a special kind of dirt so that it doesn't hurt their hooves—and, apparently, it's very expensive—so we put a whole mess of plastic down out back, and put all the dirt on the plastic. At some point, it'll go back."

The Jacob K. Javits Convention Center in New York City undergoing transformation by the Army Corps of Engineers into an alternate care facility during the COVID-19 pandemic. (*Photograph by K. C. Wilsey, Federal Emergency Management Agency; courtesy of the Army Corps of Engineers*)

At New York's Javits Center, Semonite and his team laid out rows of cubicles in long aisles, mapping the flow of patients, staff, and materials through the facilities in order to maintain a barrier between clean and dirty. "We had to figure out: When the ambulances pull up, where do they go?"

he said. "We put together a playbook for trash removal, food, dirty gowns." Each facility needed a secure area to store and dispense drugs; each cubicle needed an oxygen line, just in case. "The stress on the nurses was unbelievable—they were holding phones so that families could talk to their loved ones in the last couple of minutes," Semonite told us. "So we built what I would call a pretty nice reception area, like an airline lounge, with big overstuffed chairs and couches, TVs on the wall, that kind of stuff."

In the early months of the COVID-19 outbreak in the United States, the Army Corps of Engineers "site-assessed" eleven hundred facilities, of which seventy-four were actually built—thirty-eight by the corps itself and thirty-six by local authorities using the corps's template—for a total of thirty-one thousand bed spaces. Many of those beds were not used, in part thanks to a perverse economic disincentive. "Well, apparently, hospitals make a lot more money when a patient is in the hospital," Semonite explained. "It got to a point—and I'm not going to say anybody was greedy—but if a hospital had the option of sending them to one of our facilities or trying to keep them in the hospital, even if it meant that they jammed two or three people in a room, they decided in some places to keep them in the hospitals."

In the U.K., where the military was also tasked with turning stadiums and conference centers into what were named "Nightingale hospitals," after Florence Nightingale, the trailblazing Victorian nurse credited with modernizing the profession, healthcare is nationalized, sidestepping this particular problem. Nonetheless, even in the U.K., at some facilities, few beds were used, while, at others, patients were turned away for lack of nursing staff, leading critics to dub them white elephants. The estimated setup costs were just shy of $300 million, and, on average, each cost about $4 million to run per month—money that some said could have been better spent elsewhere.

Such high-speed repurposing and subsequent gripes about its expense are as old as quarantine itself. The historian Jane Stevens Crawshaw draws on the testimony of a Genoese friar, Father Antero, whose 1657 account of the city's lazarettos revealed that monasteries and military buildings had been requisitioned during plague outbreaks to house the sick as well as those under observation. According to Antero, pop-up structures—sheds and huts capable of housing more than a hundred people each—were also built beyond the city walls. These were temporary facilities that would be disinfected by being burned to the ground when the outbreak was over. "The system was notoriously expensive," Crawshaw writes, but, for the most part, accepted as essential. As Antero's contemporary Ludovico Antonio Muratori, the librarian to the Duke of Modena, put it, both plague and war "were 'matters of state' and states invested in fortifications against both."

"Us in the military, we're always into the reserves," agreed Semonite. "We always want to be able to make sure we have extra, so if the worst scenario happens, we're able to handle that." Semonite brought up the post-COVID fate of a basketball court the corps flipped in Westchester, just north of New York City: Could the equipment and oxygen lines simply be saved, stored, and serviced once a year at taxpayer expense, so that the facility could be redeployed in a fraction of the time and cost in the future? "We've got to learn from this, and figure out, as a nation, what is the reserve capability that is affordable," he said. "Next time, we can't afford to say something like, boy, we should have seen this coming."

Indeed, Dr. Georges Benjamin, executive director of the American Public Health Association, told us that this secondary use should be designed into the country's built environment from the ground up. "By the way, quarantine is only one need for those things," Benjamin added. "As part of our overall public health preparedness, we have to look at putting people up because of a hurricane, or floods, or a tornado—or a big infectious outbreak."

If architects considered the need for infection control, toilet facilities, electrical outlets, and other necessities when designing new stadiums, convention centers, and sports facilities, those buildings would be substantially cheaper and faster to flip in an emergency situation—not to mention much easier to return to their original function once that danger had passed. "I can tell you that a lot of work had to be done to fix and clean the New Orleans Superdome," said Benjamin, describing the aftermath of Hurricane Katrina in 2005. "But if you had built it so that it could be much more functional in an emergency situation, then you would have had less damage."

What's more, Benjamin told us, the dual-use value of such facilities ought to make them a more appealing investment for civic leaders, just as, according to Crawshaw, the provision of temporary plague hospitals was embraced in sixteenth-century Italy as "an opportunity for rulers to display their noble virtues."

In Wuhan, China, the city where COVID-19 first emerged, Yan Zhi played the role of Renaissance prince. Yan—who has been described by his alma mater, Wuhan University, as "a poet among entrepreneurs and also an entrepreneur among poets"—started out as an advertising executive before launching his own shopping mall development business, Zall Smart Commerce Group. In the past decade, he has both edited an anthology of Chinese poetry and earned a spot on *Forbes*'s list of the world's wealthiest individuals.

Like a Genoese duke, when the pandemic struck his hometown, Yan stepped up, drawing on his company's resources to transform a conference center, an exposition hall, and a gymnasium into three temporary hospitals in Wuhan, and to set up and provision a handful more elsewhere in Hubei Province. These structures were called *fangcang* shelter hospitals. The concept was similar to the U.S. Army Corps of Engineers' alternate care facilities or the U.K.'s Nightingale hospitals, but with distinctly Chinese characteristics in terms of both design and use.

After a bungled initial response, hampered by covering up the possibility of person-to-person transmission, the Wuhan authorities locked down the city and announced a bold new plan to squelch the pandemic, reported under the bullish headline: "The First Shots in the Battle Against the Coronavirus in Wuhan Have Already Been Fired!" Starting in early February, local officials went door-to-door, taking temperatures and asking about symptoms. Anyone suspected of having the virus or known to have had contact with a COVID-19 patient was sent to a *fangcang* facility. There, they were tested: those whose results were negative were sent home to shelter in place during the city's stringent seventy-six-day lockdown; those who tested positive but had only mild or no symptoms were kept at the *fangcang*, to be monitored and isolated en masse.

If their symptoms worsened, these patients would be transferred to one of the city's regular hospitals. In the meantime, these otherwise relatively healthy individuals were confined to the *fangcang*, unable to leave or even see friends and family until confirmed free of the disease. The facilities were carefully laid out to prevent cross contamination, and everyone wore masks. Nevertheless, the physician behind the concept, Wang Chen, emphasized that the *fangcang* were designed as "a community," in which otherwise isolated patients could "support each other and engage in social activities." Photos showed patients square-dancing together at six-foot intervals, as well as celebrating birthdays; the handbook put together by Yan Zhi as a freely available guide to the construction and operation of the *fangcang* includes sections on sewage management and disinfection protocols, but also a list of recommendations to "enrich the cultural life of patients" and "enhance their confidence and courage to conquer the virus."

In a text message to fellow writer Fang Fang, whose *Wuhan Diary*, initially published on the Chinese microblogging website Weibo, provided a relatively unfiltered real-time window onto the city's experience, Yan Zhi wrote that "we are going to install a lot of television sets, set up a small library area, a charging station, a fast-food area, and make sure that each patient gets at least an apple, a banana, or some other fresh fruit every day; we want the patients to feel like we care." In her subsequent diary entry, Fang Fang added that she had seen some doubts about the temporary hospitals circulating online, but, based on selfie videos shot by patients, she concluded that the "facilities there are quite good and the patients there seem to have a positive outlook."

Certainly, Chinese doctors broadly credit the *fangcang* system with allowing them to contain the pandemic. According to data published by biostatisticians from Wuhan and Harvard University, the number of new cases in Wuhan peaked on January 24, declined rapidly after the institution of centralized quarantine, and, by February 6, fell to a reproduction rate lower than one (in other words, on average, each infected person passed the virus on to fewer than one person). In late 2020, authorities reported that Wuhan had not had a single new case of COVID-19 since mid-May, and, while Europe and the United States struggled with second and third waves of the virus, Wuhan residents happily attended music festivals, pool parties, night markets, and cinemas, all without masks.

Even as Todd Semonite and his team were building alternate care facilities around the United States, some public health experts called on authorities to use them to house all positive cases and their contacts, rather than just the seriously ill. In a *New York Times* op-ed published on April 7, 2020, Dr. Harvey Fineberg, former head of the Harvard School of Public Health, and colleagues argued that using "separate designated facilities to accommodate and monitor those isolated with mild illness and those subject to quarantine" would, indeed, "require us to endure new and difficult challenges." However, they added, this approach would also offer the best outcome, in terms of fewer infections and deaths, as well as a much faster return to normality.

"We were having these same kinds of discussions at the CDC," sighed Marty Cetron. "From a technical point of view, it makes more sense to do it the way the Chinese did it. From a sociological, ideological point of view—the rugged individualism of American capitalism and democracy in contrast to the community order and tight government control in China—you're just looking at two different kinds of acceptability for those things." The same tension played out in Italy. In March 2020, a visiting delegation of Chinese doctors emphasized the importance of *fangcang* facilities for centralized isolation and quarantine; the next month, Giovanni Rezza, director of infectious diseases at Italy's national health institute, told *The New York Times* "that the government did not think that a centralized effort was 'feasible, possible, appreciated.'"

"When we were doing all this in April, no one really thought through much the quarantine piece," Semonite said. Internally, he told us, the military quickly instituted a Chinese-style policy, confining all soldiers who were positive, even if asymptomatic, to a special barracks to be monitored and isolated. “That way, we’re able to keep them kind of centralized and controlled,” he said. “But we didn’t really get asked to do any quarantine stuff for the cities.”

Such a facility would also be easier to retrofit, Semonite continued, his engineer’s mind immediately moving into problem-solving mode. “Maybe there’s two or three different types—if somebody has preexisting conditions and they’re over a certain age, maybe that quarantine room looks a little bit different.” As if anticipating the rugged individualistic resistance raised by Cetron, he wondered out loud: “I guess the question is: Is there even a way to do this in a decentralized way?”

In March 2020, Semonite had asked his research-and-development team to investigate whether portable moving containers, such as PODS, could be retrofitted as individual containment units; now he speculated as to the feasibility of designing and delivering the materials to turn an average American bedroom into a quarantine unit. “I’m just talking off the top of my head, but think about a kit,” he told us, before enthusiastically describing the kind of plastic zipper doors sold at hardware stores to seal off dust during home renovations. “It’s going to be a lot, lot simpler, obviously, than a hospital room,” Semonite said.

Nonetheless, it quickly became clear that the decentralized approach introduced some wrinkles that would be harder to solve through engineering alone. “We’re worried to a degree, I think nationally, and in the military as well,” admitted Semonite, shifting focus from the nuts and bolts of constructing quarantine to its emotional impact. “What happens when you have people who are in this status for too long, someone who doesn’t have a job, someone who’s worried about getting sick? The national suicide rate is going up.”

⟷

Before 2020, it was not easy to find someone who had firsthand experience of the rigors of quarantine. The survivors of the last mass quarantines, during the 1918–19 flu pandemic and the American Plan, had almost all passed away. Since then, hundreds of thousands of people have been quarantined—thousands of children during the polio outbreaks in the early 1950s, before a vaccine was developed; thousands of people in Toronto and Taiwan for SARS in 2003; and entire urban neighborhoods and villages for Ebola in the past decade. Still, quarantine has been far from a universal experience: in 2019, only an unlucky minority had endured its trials.

The physician Patrick LaRochelle joined that select group when, without wearing gloves, he unwittingly treated an Ebola patient while working in a hospital in the Democratic Republic of the Congo. LaRochelle is a sweet-faced, bespectacled young doctor; he and his wife are committed Christians and chose to serve in the DRC as part of a missionary organization.

On the fateful day in late December 2018, LaRochelle was finishing up his shift, about to head home, when he heard that a woman with low oxygen levels and a dead fetus in her womb had just been transferred to the intensive care unit. Ebola cases had been on the rise in the neighboring town, forty miles southwest, but, although the hospital had spent the past month on high alert, no infected patients had yet presented themselves. "We were experiencing what you might call vigilance fatigue," he admitted. LaRochelle got out his stethoscope, listened to her heart and lungs, then asked where the patient was from. The answer—a village where cases were starting to multiply—made him notice that the patient's eyes were also red and that she was oozing blood from the site where nurses had tried to put in an IV. She died two hours later—and blood tests confirmed that she was infected with Ebola.

"What followed next is a blur," LaRochelle said, showing us a heartbreaking photo of his wife and two young children miming hugs from six feet apart, their arms outstretched as if they are falling toward him. Given the unstable situation in the region, LaRochelle was quickly separated from his family and evacuated by the State Department. At that time, the new National Quarantine Unit was still under construction, so Ted Cieslak and the University of Nebraska Medical team were tasked with housing and monitoring LaRochelle in their biocontainment unit for two weeks.

Although LaRochelle knew that his exposure was unlikely to result in infection, there was still a part of him that couldn't help but wonder if he had touched his wife and children for the last time. Still, he told us, "being a bit of an introvert, the time was actually restful and relaxing." Ahead of his arrival, nurses had printed and framed photos that LaRochelle's wife, Anna, had emailed, and stocked the room with novels and magazines. He took advantage of the free time, the treadmill, and the stationary bike to try to get in shape for the first time in a decade—a goal that was undermined by the doctors and nurses who would stop by to drop off Omaha barbecue and cookies.

When we asked LaRochelle how his experience of quarantine had affected him, he told us that it had made him much more aware of how traumatic compulsory isolation can be. Not because it felt that way to him: on the contrary, he told us, he realized that, although Americans are perhaps the people most likely to protest the idea of mandatory quarantine, that same individualism almost uniquely prepares someone to endure and even enjoy the experience. After all, quarantine, in a different context, can just look like a two-week Netflix binge. By contrast, LaRochelle said, he later asked his children's Congolese nanny, Denise, how she might feel if she were kept isolated for a few days in order to ensure that she didn't have Ebola. "A look of horror came over her face," he said. As he spoke with Denise, it became clear that it was the prospect of separation from her family and community that terrified her, much more than the idea of limiting her own freedom in order to protect them, or even the possibility of having Ebola.

LaRochelle's Congolese colleague Patrick Ucama confirmed this assessment. "It's not good to leave a member of family alone somewhere," Ucama told us. "That is a basic idea to our culture." As a result, he said, when he tried to detain patients who seemed suspicious for Ebola, "for every person I sent to isolation, only one accepts, and nine others refuse and go home."

Ucama's experience has been shared by public health officials across West and Central Africa, and is seen as a significant factor in the outbreak's spread. In Sierra Leone in 2014–15, an estimated one-third of patients hid from contact tracers and refused to report to treatment centers; across the region, dozens of patients who had tested positive for Ebola fled the treatment centers, sometimes with the help of their families. With Western medicine unable to offer a cure, traditional healers and home-based care seemed more appealing than suffering in enforced isolation, surrounded by strangers in hazmat suits. "I think that things that seem relatively insignificant, like the suits you wear, can have a huge effect on people," LaRochelle told us.

As we heard this, we were reminded of our visit, a year earlier, to the Royal Free Hospital High Level Isolation Unit in London, where isolation has been designed quite differently. "We've been mocked for having this system here," said Dr. Sir Mike Jacobs, the unit's clinical director, when he showed us one of the hospital's two Trexler isolators. "No one else has adopted this system, but we're very happy with it."

At Nebraska, and pretty much everywhere else in the developed world, biocontainment care is achieved by putting a patient in a negative air-pressure room and dressing staff in layers of personal protective equipment. In the U.K., by contrast, Ebola patients are put inside their own negative pressure, transparent plastic tent; in order to look after the patient, doctors and nurses insert themselves into half-suits built into the PVC walls. The system was designed by Dr. Philip Trexler in Indiana, in the 1950s, and was originally envisioned as a sterile environment in which germ-free animals could be bred for medical research. "Trex," as he preferred to be called, soon dreamed of installing his isolators in hospital operating rooms across the United States in order to prevent postsurgery bacterial infections, but the proliferation of new antibiotics during the 1950s and '60s seemed to provide cheaper and easier infection control. Instead, he moved to the U.K.'s Royal Veterinary College in 1966, to lead a program intended to breed pathogen-free pigs for industrial production. When the first cases of Ebola, Lassa, and Marburg viral hemorrhagic fever emerged from Africa in the 1970s, an adapted Trexler unit was trialed for infectious disease isolation. It caught on.

While Nicola crawled inside the Trexler to experience the patient point of view, Mike Jacobs demonstrated how he would provide care. "You put this air jacket on, first of all, to keep you cool—it just blows cold air over you," he explained. Then he ducked down, put his arms in the sleeves, pushed his head into the billowing shower curtain–like material, and wiggled it down around him as if putting on a sweater, only to pop up on the inside of the tent—but entirely enclosed in a layer of plastic. "You're not standing back from it, you're actually inside—but, obviously, on the

Nicola Twilley sits inside the Trexler isolation unit and talks to Dr. Sir Mike Jacobs through its plastic wall. *(Photograph by Geoff Manaugh)*

Looking into a Trexler isolation unit at the Royal Free Hospital, London. *(Photograph by Geoff Manaugh)*

outside, if you see what I mean," said Jacobs. "You've invaginated yourself into it."

The advantages of the Trexler system were immediately clear, especially after having donned PPE ourselves with Luigi Bertinato in Venice. The multiple layers of plastic and Tyvek gowns, hoods, gloves, boots, apron, and face protection take a full ten minutes to put on and take off correctly, and quickly become extremely hot and sweaty, which is why nurses like Kaci Hickox were supposed to work for only forty minutes at a time while treating Ebola patients in Sierra Leone. In the Royal Free isolation unit, nurses work twelve-hour shifts, with no danger that they'll accidentally infect themselves while adjusting steamed-up goggles or peeling off their Tyvek suits. Indeed, Jacobs said, anyone can "invaginate" themselves into the tent with a minimum of instruction, meaning that other medical specialists without infectious disease training and even family members can safely visit the patient in person.

"We are absolutely clear that, for a low-volume system like ours, with one case every two or three years, maintaining the ability to safely look after patients is so much easier when you use a systems solution to biocontainment that doesn't require intensive training and maintaining that training," Jacobs told us. From inside the Trexler, it was possible to hear him clearly, and, more important, see his face. Although the tent was spacious and light-filled, it was easy to imagine it might feel claustrophobic after a while. "But I think there are some advantages that are slightly counterintuitive," said Jacobs, "because you have to set it against wearing PPE, which dehumanizes you enormously. Although patients are confined in terms of the space, there's actually much more real human contact than if people were coming in in space suits all the time, where you can't see their faces. In a way, that is much more frightening."

Recently, Dr. Richard Kojan, a Congolese ICU physician frustrated by treating Ebola patients in PPE, visited the Royal Free, then designed a cheap pop-up Trexler-like tent called the CUBE, or the biosecure emergency care unit. In December 2018, an eighteen-CUBE center was set up in Beni, a few hours' drive south from Patrick LaRochelle's hospital in the Democratic Republic of the Congo. In an Associated Press video, a doctor inserts himself into the half-suit and leans over a woman with Ebola to give her a sonogram; next door, a husband visits with his sick wife, smiling and chatting through the plastic. "It's much more sociable, if I can put it that way," said the doctor. "When she looks at me and smiles, I feel that we are together again and that she will quickly return home," the husband agreed. "Experimenting, being creative about how you quarantine, based on the situation, is huge," LaRochelle told us. "How can quarantine protect the community even while reminding those who are isolated that they're actually part of our community?"

Intriguingly, the other situation to which Trexler isolators have been adapted is to house immunodeficient children, such as David Vetter, the "Boy in the Bubble," whose life story inspired a 1976 made-for-TV movie starring John Travolta. When Vetter died at the age of twelve, he was the first human being to have spent his entire life in a state of medical isolation, quarantined from the world while awaiting a breakthrough in treatment.

Researchers found that the experience had left him with a deeply distorted sense of space and time. Vetter's spatial awareness was almost nonexistent, and "the idea of limitless space confused and scared him," according to the historian of medicine Robert Kirk. Meanwhile, "as time determined all activities that happened about him, David had developed a highly acute sense of time," Kirk wrote. "It was time, not space, by which David had learned to orientate his world."

⟷

Fear is, unsurprisingly, a common reaction to quarantine: fear of isolation; fear of disease; fear, fundamentally, of the unknown. What little historical testimony that survives—perhaps because of its inherent and seemingly unheroic passivity, the enforced lacuna of lazaretto life is often elided from memoir and travel narratives—shows quarantine to have been, in general, "a most unloved institution." That phrase comes from David Barnes, who has tracked down letters and diary entries written from the Philadelphia Lazaretto in the course of documenting the building's history. "Many refer to complete and utter boredom, and to impatience—sometimes to fear of getting sick," he told us.

In his study of Mediterranean quarantine, the historian Alex Chase-Levenson quotes a travel memoir by an English woman, Mrs. Griffith. "I never thought of the plague till now," Mrs. Griffith wrote, complaining that her admission to the Maltese lazaretto had the ironic effect of making her "almost terrified" of becoming sick. Charles Dickens begins his 1857 novel *Little Dorrit* with a scene set in quarantine at Marseille, in which the otherwise jovial Mr. Meagles laments, "I have had the plague continually, ever since I have been here. I am like a sane man shut up in a madhouse; I can't stand the suspicion of the thing . . . I have been waking up, night after night, and saying, now I have got it, now it has developed itself, now I am in for it."

For many European travelers, time spent in quarantine was a deeply unsettling experience: suddenly, *they* were the outsiders, potentially rendered too dangerous and dirty by their travels abroad to be allowed back home, their readmission to civilized society made uncertain and contingent. Chase-Levenson quotes the quarantine reminiscences of an officer of the East India Company, David Lester

Richardson, who wrote that it was a most "curious feeling" to be among "such marked and suspected people, and to know that strangers would be horrified at our touch."

In the twenty-first century, the COVID-19 lockdown initially proved equally fearful. In Wuhan, Fang Fang describes the time surrounding the city's quarantine order as "five terrifying days," during which "most people in Wuhan were in a state of utter panic," while the virus roamed the city "like an evil spirit, appearing whenever and wherever it pleases, terrorizing the people."

Such a fever pitch can hardly be maintained for the duration of quarantine, and, before long, the experience typically transitions into a minor-key medley of melancholy emotions: anxiety, loneliness, stress, and sadness, all overlaid with a pervasive, unshakable tedium. As Albert Camus puts it in his novel *The Plague*, after several weeks of enforced lockdown, "The furious revolt of the first weeks had given place to a vast despondency."

During 2020, such boredom and despondency were evident in the fact that Britons reportedly spent nearly half of their waking hours watching Netflix during lockdown, alcohol sales in the United States increased by 27 percent in the first three months of the pandemic, and hundreds of people around the world spent thousands of dollars on "flights to nowhere"—circular journeys, complete with inflight catering, for "those who want to spread their wings" in spite of travel bans.

The options for alleviating boredom available to the quarantined of Genoa in the 1650s were perhaps even less appealing: according to Jane Stevens Crawshaw, "Father Antero recommended that women could be kept busy with sewing and mending or making shirts, swaddling bands for infants and ecclesiastical decorations, as well as by spiritual devotions." These activities were, of course, remarkably similar to the entertainment options already available outside the constraints of medical detention: one of the ironies of quarantine is that it renders everyday life unbearable by making it feel involuntary—even inescapable. St. Charles Borromeo, in his role as archbishop of Milan in the 1500s, admonished his flock that, rather than expecting to enjoy quarantine, they should instead return to the practice's conceptual roots. "Each person should prepare themselves to use this time well," he advised, "and treat every day of this quarantine as being like the holy time of Lent."

In quarantine in the 1830s at Zemun (then Semlin) on the Austro-Hungarian sanitary border, the artist Francis Hervé wrote that his traveling companion, lacking such a spiritual crutch, occupied the first day by writing letters and the second by tallying his travel expenses. "But on the third day, and ever after, he had no resource to drive away the fidgets, and moaned and groaned incessantly," wrote Hervé. "He would go to bed at eight o'clock, to kill some portion of time; but even that expedient did not avail him, as he found he could not sleep when he retired so early."

"Boredom is just a signal," explained Erin Westgate, a psychologist at the University of Florida who studies the emotion. "It alerts us that what we're doing isn't meaningful or we're not engaged in it well." The ways in which Americans have responded to that signal during the COVID-19 lockdown varied widely: doom-scrolling through social media, picking fights with quarantine rule-breakers and mask-avoiders, repeatedly opening the fridge door for inspiration, or just pouring a drink. (Alcohol, Westgate's research has shown, does effectively suppress boredom, "temporarily, at least.")

Others have found more virtuous coping strategies: sewing masks and delivering food for neighbors, baking sourdough, learning languages, producing Zoom operas, or simply immersing themselves in the delights of nature or their neighborhood. History is full of examples of such enviable productivity: the French poet and statesman Alphonse de Lamartine claimed to have written the Serbian section of his acclaimed 1835 travelogue, *Voyage en Orient*, in quarantine at Zemun; not to be outdone, the British politician Benjamin Disraeli produced drafts of two novels while confined to the lazaretto at Malta during the summer of 1830. Mr. Meagles, in Dickens's *Little Dorrit*, ends up nostalgic for his sojourn in the Marseille lazaretto. "'But, Lord bless me!' cried Mr. Meagles, rubbing his hands with a relish, 'it was an uncommonly pleasant thing being in quarantine, wasn't it? Do you know, I have often wished myself back again? We were a capital party.'"

Indeed, the 1800s witnessed the brief efflorescence of a new literary subgenre, in which the leisurely, necessarily cosmopolitan environment of the lazaretto served as the setting for the discovery of shared interests and even the blossoming of romance. ("Love in a Lazzaret," a syrupy short story published in 1830 in *The Knickerbocker* magazine, is a classic of the genre, in which Yankee traveler Delano develops "the most incontestible [*sic*] symptoms"—of love, rather than cholera—for a fellow medical detainee.) As the literary scholar Kelly Bezio writes, "The quarantine narrative worked to depict the common humanity of all nationalities as a source of comfort"—a disappointingly rare perspective on quarantine's power to unify, as well as to separate. Two centuries later, the COVID-19 pandemic stimulated its own rise in quarantine-themed erotica, with steamy tales involving hand sanitizer, Zoom voyeurism, and strangers forced into lockdown together.

Westgate's own research—including something she referred to as "the jerk study," in which she forced people to watch an incredibly boring video of a rock, then gave them the option to take away someone else's money by pressing a button, for no other reason than to have something to do—presents compelling evidence that bored people with only bad options will behave in ways that harm either themselves or others (or, often, both). Conversely, Westgate told us, if those same bored people were also given the option to give someone else money, almost everyone will. "But, if you don't give them that option, they kind of look around and say, well, there's nothing else to do, so I guess I'll do the bad thing," she said.

Extrapolated to the tedium of quarantine, she continued, this research suggests that authorities might want to think through not only the logistics of quarantine in advance but also the experience: the *why*, in addition to the *where*, *what*, and *how*. "It's not enough to simply say: everybody go home and stay there," Westgate said. "You want to try to give people something meaningful to do during that period. It doesn't have to be a *happy* experience; it just has to be psychologically rich."

⟷

"I felt like an expert when I talked to you before," said Marty Cetron, "but now I realize how much I had to learn in the actual use of quarantine." Before COVID-19, Cetron had spent more time reading, thinking, and talking about quarantine than any other public health official we met. No one should have been better prepared for its challenges and compromises. Yet, when we spoke to him a few months into the coronavirus pandemic, even his laughter sounded anguished. "Trying to navigate an antiquated, scary public health tool with nonexperts in a real-time setting of existential threat during an election year with a really divided American populace . . . It's just—I can't even tell you how overwhelming," he told us.

For all the technological, legal, ethical, and architectural ingenuity that had been applied to updating quarantine, when it came time to try to deploy it in order to contain the first global pandemic of the twenty-first century, the CDC and public health authorities across the country, for the most part, utterly failed. The virus itself did not make the challenge easy—its airborne mode of transmission and asymptomatic contagiousness made COVID-19 both "fast-moving and insidious," as Cetron put it. The CDC's own widely publicized failures to implement fast and accurate testing allowed the virus to spread through the community, to the point where the only options for containment were large-scale quarantines or lockdowns.

Many of the reasons that the United States, in particular, struggled to control the spread of COVID-19 were quite predictable—indeed, most of them had already been outlined by public health scholars. Twenty years ago, David Fidler, a cybersecurity and global health expert at the Council on Foreign Relations, detailed a few of the many ways in which the United States is especially vulnerable to a virulent microbe. For starters, he pointed out, a federal structure in which public health powers reside at the local level will, by design, respond to a pandemic in a fragmented and uncoordinated manner. "A legal system that emphasizes the protection of individual rights and restricts governmental powers to impinge on such rights," Fidler added, is also one in which quarantine and other social distancing measures are destined to fail. "In such a system, the citizenry is always wary and skeptical of governmental incursions on its rights, creating a climate of distrust that works against governmental efforts to contain an epidemic."

The United States has neglected its public health infrastructure for decades, Fidler wrote. "People forget that the 'rule of law' goes beyond, and must go beyond, merely having legal powers on the books. The legal power to act in the public good must be supported by resources, personnel, training, and equipment to undertake effectively the legal authority that exists."

Despite Marty Cetron's and Kaci Hickox's efforts; Ted Cieslak's National Quarantine Unit; and countless conferences, simulations, and tabletop exercises, not enough changed over the two decades following the publication of Fidler's paper. Matthew Penn, director of the public health law program at the CDC, confirmed our suspicions that very few experts had considered all, or even any, of the steps when it came to quarantine. "Where can quarantines take place?" Penn rhetorically asked a roomful of judges and court officials, as part of a national summit on pandemic preparedness held in May 2019. "People occupy space, they are in places, but when it comes to planning, logistics, how that happens—our laws are silent on these."

Penn, whose enthusiasm for untangling the intertwined arcana of local, state, and federal case law was evident, spent the next twenty minutes asking question after unanswered question, with only the occasional pause for dramatic effect. "What are the penalties for noncompliance with quarantine?" he asked. After all, a person under quarantine might be supposed to stay at home, but enforcement is another matter. "Let's say the person under quarantine order does leave the house," Penn continued. "Is it a fine? Is there an arrest? Who does the arrest? I can tell you, the law is sometimes not going to say."

Before he joined the CDC, Penn served as a staff attorney for the South Carolina Department of Health, a state that does have a statute, passed in 1954, compelling sheriffs, constables, and police officers to assist in carrying out quarantine orders. As he spoke to law enforcement in the course of his job, Penn said, he would ask them about this statute. Few of them had heard of it; most of them told him they had no intention of doing any such thing. "One of the things that I heard across South Carolina is: we're just not going to get involved," Penn said.

To anyone who heard Penn's talk, the widespread refusal of sheriffs across the United States to enforce bans on large gatherings during COVID-19 should not have come as a surprise. It felt equally inevitable, to anyone who had read Fidler's paper, that resistance to public health orders—for example, refusing to protect both oneself and others by wearing a mask—would erroneously come to symbolize a defense of American constitutional freedoms.

"It is striking how little global society was able to take advantage of the historical examples, that had been studied in great detail, and to actually apply those lessons," Marty Cetron observed. His own research into the efficacy and timing of public health interventions during the 1918 flu had not only shaped the U.S. pandemic playbook but also formed the basis of the WHO's own response framework. Other countries, such as South Korea, used the CDC's plan, and largely succeeded in containing the virus, albeit at huge cost.

"I should have paid more attention to the politics and context of 1918 to realize that some of the ingredients of success and failure were going to be dependent on leadership and communication and coordination," Cetron told us. During COVID-19, the U.S. public health response became politicized almost immediately, in a country where trust in scientific expertise was already deeply eroded. "We trained I don't know how many other countries in how to do all of this and the importance of it," said Cetron. "And here we are: everyone else in the world that is succeeding is using the power of this strategy and the intellectual energy of the U.S. government, and"—he shook his head—"our own leadership is just thumbing their noses."

Another blind spot, Cetron confessed, lay in considering quarantine and other distancing measures simply as public health tools—something to slow the spread of a disease in certain situations—rather than as a lived experience that would somehow have to be implemented, enforced, and endured for months on end. This was a gap we had noticed again and again: at multiple pandemic simulations, we saw participants invoke quarantine, then move on to the next challenge, assuming that the measure had worked as expected. "Most exercises that I've been involved in have a heavy emphasis on the opening moves, and then they jump quickly to the closing stage," Cetron agreed. "Nobody appreciates what the middle game is going to look like, but that's the hard game."

In reality, he told us, the restrictions that can control spread and reduce deaths in a pandemic such as COVID-19 have to be implemented earlier and be more stringent than seems reasonable, and they have to be sustained for a period that outstrips everyone's patience.

"We just didn't have the grit for it in the United States," said Cetron, with a deep sigh. "It's quite humbling."

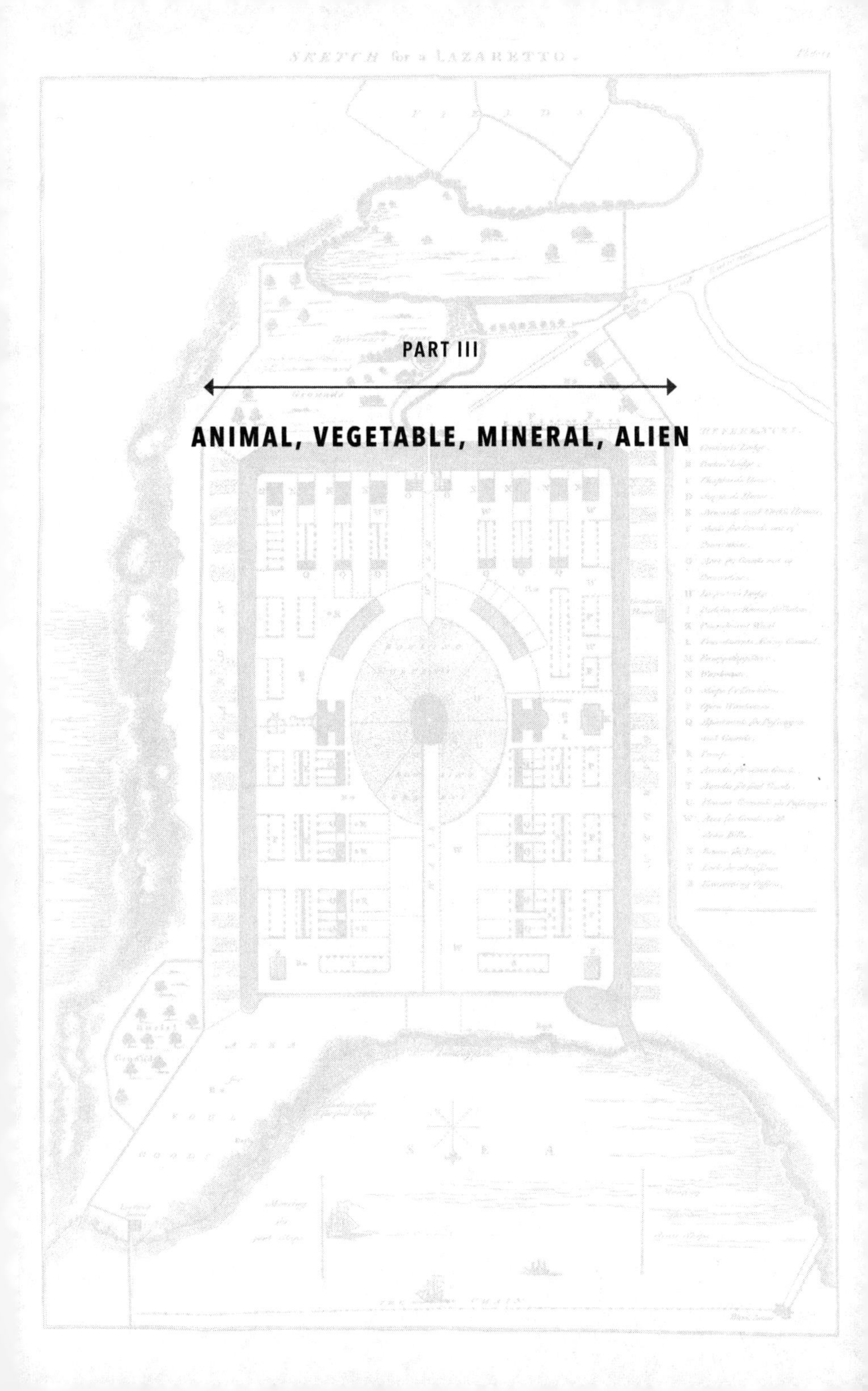

PART III

ANIMAL, VEGETABLE, MINERAL, ALIEN

6

Biology at the Border

Most spiders are lonely creatures, but not the ones kept by the evolutionary biologist Noa Pinter-Wollman in her Los Angeles lab. These small, light-brown, velvet-coated African social spiders live in intricate silk architectures that house several hundred individuals. The colony's kinship ties are even more tightly woven, with virgin "auntie" spiders offering themselves up as a sacrificial food source for the next generation.

Like humans, social insects reap invaluable benefits from living together: by working as a group, Pinter-Wollman's spiders can capture prey ten times their own bodyweight, as well as weave denser, stronger webs that offer better shelter and protection from hungry birds. As with human communities, however, the downside of being social is that close proximity to others increases the risk of catching an infectious disease—a fate that becomes more likely in wet conditions, when fungal pathogens can spread rapidly through the nest, wiping out an entire colony in days.

As Pinter-Wollman studied her spiders, each anointed with a dab of neon paint for ease of identification, she realized that, as a group, they adjusted their behavior according to the risk of disease, engineering a careful balance between acquiring food and avoiding infection. Bold spiders predominated in the meager hunting grounds of a dry container, while their cautious counterparts prevailed in circumstances of damp abundance. In other words, their response to pathogenic risk looks a lot like spider quarantine.

Pinter-Wollman is one of just a handful of researchers who, in the past couple of decades, have begun to explore how other, nonhuman creatures whose lives take place in intertwined communities—ants, honeybees, termites, and spiders—design their architecture and modify their movements in order to reduce their exposure to disease.

Underlying this research is an often-unspoken hypothesis: that all social species have been shaped by these difficult trade-offs for millennia, and that thus, perhaps, there are some deep evolutionary laws underlying social distancing and quarantine. Pinter-Wollman's spiders were in the same boat (or, in this case, plastic bucket) as medieval Venetian city officials who had to weigh the loss of lucrative trade with the East against the ravages of the Black Death, or, more recently, U.S. state governors negotiating calls to open the economy while also flattening the curve during COVID-19. Faced with an uncertain but existential risk, African social spiders and political officials alike introduce a little hesitation and delay—even a curb on some of their movements—in order to try to have the best of both worlds.

Early experiments have shown that some infection-control behaviors are certainly shared across species: honeybees that have been attacked by their most devastating pest, the *Varroa destructor* mite, often decide not to return to the hive, self-isolating so as not to spread disease to their nestmates; similarly, when forager garden ants detect the presence of deadly fungus spores while searching for food, they subsequently keep their distance from their indoor colleagues, increasing the time they spend outside the nest. Even vampire bats have been found to reduce their interactions with fellow bats when sick.

In a slightly less relatable move, termites often cannibalize their young during an outbreak. Nina Fefferman, a mathematician and epidemiologist who frequently collaborates with Pinter-Wollman, likened termite larvae-eating to school closures in humans: a way of neutralizing risk from the demographic that is both most susceptible to infection and most likely to transmit it.

Already, looking at bugs has taught humans a trick or two, which is perhaps not surprising: Pinter-Wollman pointed out that ants have been builders for much longer than humans. "These structures have been through selection pressures for thousands and thousands of years," she said. For example, the way in which ants handle bottlenecks has led to the insight that, for humans, placing a pole in front of a building's emergency exit will actually speed up evacuation, as people flow around either side.

More recently, Fefferman has adapted for human use one of Pinter-Wollman's observations about disease transmission in insect societies. "In both bees and harvester ants, the progression of work is very, very regimented," Pinter-Wollman explained to us. Bees start life by taking care of the queen and the brood, and, as they age, they progress to outdoor activities; the last job harvester ants undertake before they die is working in the colony's refuse pile. "The way that labor is divided means that the individuals that are most likely to be infected are spatially farther from the inner nest," Pinter-Wollman said. Inspired by this insight, Fefferman has developed a cohort-based formula for companies to restructure their workforces in such a way as to maintain productivity while reducing infection rates during a pandemic.

Of course, as Pinter-Wollman was quick to remind us, we don't yet know for sure whether insects evolved these behaviors and social structures in order to protect against pathogens or for some other reason, with disease containment simply a beneficial side effect. Before the coronavirus shut down her lab, she was about to begin studying another open question: whether ants alter the structure of their nest in response to local disease conditions. In a pathogen-rich environment, might ants build smaller entrance chambers that connect to fewer rooms, deliberately sacrificing a degree of foraging efficiency to slow down the rate at which an infection would spread through the nest? Certainly, in some cases, other species have been known to decide that the cost of quarantine-type behavior simply isn't worth it: gorillas in the Congo are more likely to die from Ebola if they live in groups, but, at the same time, females and young gorillas are even more likely to die from other causes if they go solo—so they don't. For some animals, quarantine and isolation are more dangerous than the disease.

When it comes to diseases like Ebola, it's not just the response that has the potential to be shared across species, but the pathogen, too. In fact, zoonoses, or diseases that have jumped to humans from other animals, have caused many of our worst plagues, as well as most of the infectious diseases that have emerged over the past fifty years. Historically, scientists think we have cattle to thank for measles and possibly also for tuberculosis. Pigs are likely the original host for mumps, while crustaceans play that role for cholera. Humans pick up anthrax from herbivores such as sheep or goats. And, just as with *Yersinia pestis*, the bacteria responsible for the Black Death, the smallpox virus's original home was in rodents. (Even today, these forgotten diseases make the leap from animal to human: in 2009, a New Hampshire woman was infected by anthrax at her community drum circle after inhaling spores aerosolized from a Malian cowhide drumhead, and there are still a handful of cases of the plague in humans each year, often picked up from pets that have sniffed around the carcasses of infected ferrets or prairie dogs.)

Many zoonotic diseases are transmitted by vectors—mosquitoes, ticks, or fleas—while others travel by air or in blood. Most, like COVID-19, originate in wild animals, but many, like the encephalitic Nipah virus or highly pathogenic avian flu, infect livestock as an intermediary host. The 1918 flu, which was the most severe pandemic in recent history until COVID-19, was avian in origin, although many scientists believe that domestic pigs acted as the "mixing vessel" in between, facilitating the virus's leap into humans. Some of these zoonoses will undoubtedly have the qualities necessary to cause the next global pandemic: as human settlement continues to encroach on previously undisturbed landscapes, and as climate change causes species' ranges to shift, experts agree that COVID-19 is just the first of many zoonotic outbreaks with the potential to shut down life as we know it.

Unlocking the rationale behind quarantine, as well as its implications, requires looking not just across disciplines but across species, too. As we've seen, quarantine is a matter for public health officials and doctors, but also for architects, civil rights lawyers, prison reformers, and postal historians. Now it's time to cast an even wider net. What does quarantine look like across the Aristotelian system's three kingdoms: animal, vegetable, mineral—and beyond?

⟷

In 2002, U.S. Navy SEALs scoured underground complexes in eastern Afghanistan for any shred of intelligence about al-Qaeda's network and operations. According to the retired four-star general Richard B. Myers, who served as chairman of the Joint Chiefs of Staff at the time, American troops retrieved dozens of notebooks and document caches from these missions. One in particular stood out. In it was a handwritten table, its five columns filled with penciled scribbles listing sixteen different pathogens, as well as their incubation period, mortality, and route of transmission.

At the top was pneumonic plague, a disease caused when the bacterium responsible for the Black Death is inhaled, with a fatality rate of between 35 and 100 percent. Beneath it were a handful of equally familiar names from outbreaks past, such as cholera and anthrax. What intrigued Myers, however, was that the majority of the pathogens on the list targeted plants and animals: rice blast, foot-and-mouth disease, "fowl plague" (otherwise known as avian influenza), hog cholera, and stem rust of cereals.

U.S. intelligence services already suspected that al-Qaeda was interested in bioweapons, but this notebook added weight to the idea that, as Myers put it, "they were indeed going about it." Later that year, he said, another intelligence source reported that a group of al-Qaeda members who had fled Afghanistan had ended up in the mountains of northeastern Iraq, where they were testing various pathogens on dogs and goats.

"To my knowledge, they've never gotten to the point where it was of use for them in the battlefield context," Myers told us. "But since al-Qaeda, as we found out with the World Trade Center in New York City, never quite give up on an idea, it's not something you can just dismiss and say, that was then, and they're probably not thinking about that now." In fact, he cautioned in his gravelly baritone, "I think there's other, probably classified information that would tell you that's *not* the case—but I'm not privy to all that or privy to talk about it."

The likelihood of a biowarfare attack that targets U.S.

livestock or crops is subject to debate in national security circles. On the one hand, such assaults were vanishingly rare even before they were outlawed under the Biological Weapons Convention, which took effect in 1975. In Operation Vegetarian, the British manufactured five million "cattle-cakes" laced with anthrax spores to drop on German grazing lands during World War II; the United States developed stockpiles of stem rust and rice blast intended to target the Soviet wheat crop and Japanese rice fields. However, only the Japanese are known to have actually used animal and plant pathogens, in Mongolia in 1940. (That uncertainty points to one of biological warfare's pros—or cons, depending on your perspective: unless it involves a novel, lab-developed pathogen, it's hard to prove that a disease outbreak is an intentional attack, as opposed to an accidental human or even natural introduction.)

Nonetheless, and despite the global ban, research has continued. In 2014, a dusty Dell laptop retrieved from an ISIS hideout in northern Syria—the "laptop of doom," as it was later dubbed by *Foreign Policy*—contained detailed instructions for the production and dispersal of biological weapons, as well as a fatwa endorsing their use. In the United States, the Defense Advanced Research Projects Agency (DARPA) recently launched a program called "Insect Allies" that aims to create an army of insects that could inject customizable gene-editing viruses into crops—its manager, Blake Bextine, claims the initiative's intent is to help protect the nation's wheat and corn, but the technology could obviously be used offensively. In 2010, the bipartisan Commission on the Prevention of Weapons of Mass Destruction Proliferation and Terrorism concluded that a biological attack in the United States was much more likely than a nuclear attack—and awarded the country an F grade for its readiness.

Agriculture is simply a "soft target," as Myers put it: our farms are underprotected and the relevant pathogens are not particularly difficult or expensive to manufacture and deploy. (Foot-and-mouth disease is so easily spread that a would-be bioterrorist could wipe out entire herds by dropping an infected Kleenex or rag in a field—paper and cloth are, to use John Howard's terminology, particularly "susceptible matter.")

As a sector, farming is also highly concentrated: three-quarters of the United States' vegetables are grown in just three states, while 2 percent of the nation's feedlots supply three-quarters of its beef. What's more, much of what we grow is genetically identical. Four companies sell more than 80 percent of all seed globally, and, though they each offer different hybrid varieties, many of those seeds still share the same DNA. According to the U.S. Department of Agriculture (USDA), just five bulls are responsible for a quarter of the genetic material in the country's entire Holstein herd; one bull, the splendidly named Pawnee Farm Arlinda Chief, contributed nearly 14 percent. Monocultures are exceptionally vulnerable to disease: their uniform makeup represents an all-you-can-eat buffet for pests and pathogens.

As the former secretary of health and human services Tommy Thompson declared in 2004: "For the life of me, I cannot understand why the terrorists have not attacked our food supply because it is so easy to do."

⟷

"We were looking at agroterrorism well in advance of 9/11," boasted Ron Trewyn, a former cancer researcher who led the effort to bring the United States' new, fortresslike federal biosecurity lab to Manhattan, Kansas. (The "Little Apple" is the prairie town's registered trademark.) In 1999, as vice-provost for research at Kansas State University, Trewyn worked with new faculty members Nancy and Jerry Jaax, the military veterinarians whose exposure to a strain of Ebola while working with monkeys in a quarantine facility in Reston, Virginia, provided much of the narrative tension in Richard Preston's *The Hot Zone*.

We met Trewyn in his basement office in Anderson Hall, the former Practical Agriculture Building, a Gothic castle complete with bell tower that loomed out of the murky gray sky on a deserted, postsemester campus. One of the Great Plains' notorious summer thunderstorms sent sudden sheets of water slashing down his plexiglass-reinforced window. "They have now fixed this a little bit," Trewyn said, laughing and gesturing to the newly sealed window, the paintwork around it stained where traces of water had infiltrated the building. "Anyhow, I digress: Jerry was actually running the global biodefense treaty office for the army, so he knew that there were people still working on bioweapons and knew that some of them were targeting agriculture. That gave us a little bit of a head start on things."

To diagnose deadly livestock diseases, as well as to develop treatments and vaccines for them, researchers need to work with them in the lab—but, as the Jaaxes' experience shows, the risk of an accidental exposure or release is very real. Foot-and-mouth disease, in particular, is so infectious that the live virus cannot be brought to the U.S. mainland without written permission from the secretary of agriculture. The only place authorized to work with it is Plum Island Animal Disease Center, built on a low-lying islet about the size of Central Park, eight miles off the Connecticut coast. ("Sounds charming," as Hannibal Lecter, the homicidal antihero in *The Silence of the Lambs*, murmurs when offered the possibility of a vacation there.)

But Plum Island opened in 1954, and its facilities are not

only outdated; they are also not certified to handle pathogens requiring the highest level of containment: Biosafety Level 4. BSL-4 microbes are "dangerous and exotic, posing a high risk of aerosol-transmitted infections," according to the CDC. Typically, they can infect both animals and humans, and have no known treatment or vaccination: examples include Ebola and other more recently emerged hemorrhagic infectious diseases such as Nipah and Hendra. Only three facilities in the world are currently equipped to handle large animals at the BSL-4 level, meaning that, during an outbreak, U.S. researchers would have to beg their Canadian, Australian, or German counterparts for lab space.

Following 9/11, the anthrax postal attacks, and Dick Myers's alarming discoveries in Afghanistan and northern Iraq, the Department of Homeland Security successfully made the case that it was time for the United States to have a BSL-4 large-animal facility of its own. They considered upgrading Plum Island—an option that proved both expensive, due to the need to transport building materials by boat, and unpopular with its Long Island and Connecticut neighbors—before conducting a national search for a new site. By that point, Trewyn and the Jaaxes had already built a BSL-3 lab on campus. The Biosecurity Research Institute opened its doors in 2008; its existence helped make the case that Manhattan, Kansas, in the middle of America's agricultural heartland, was poised to become what Trewyn, echoing the former senator Tom Daschle, has called "the Silicon Valley of Biodefense." A year later, the Department of Homeland Security announced that the new National Bio and Agro-Defense Facility (NBAF) would be built right next door.

Admittedly, some critics did question the wisdom of situating a lab designed to work with the world's most devastating large-animal diseases in a state where cows outnumber people by more than two to one. One in ten of America's cows reside within a two-hundred-mile radius of Manhattan, Kansas; if foot-and-mouth were accidentally released, estimates show it could easily infect herds in the surrounding states—nearly half of the nation's cattle—causing up to $50 billion in damage. The probability of such an escape over NBAF's fifty-year projected life span was estimated to be an astonishing 70 percent, in a report prepared for the National Academy of Sciences by the microbiologist Ronald Atlas.

Another report, this time from the Government Accountability Office, concluded that the Department of Homeland Security did not actually have sufficient evidence to conclude that foot-and-mouth could be safely contained on the mainland. Meanwhile, the Texas Biological and Agro-Defense Consortium, whose preferred site in San Antonio finished as runner-up, immediately filed suit with a fifty-page list of complaints arguing that the choice was political. In claim 103, it added that anyone familiar with *The Wizard of Oz* should be aware of Kansas's reputation for dangerous tornadoes. (The suit was dismissed without prejudice a few months later.)

In response, the Department of Homeland Security "hardened" the design to resist a Category 5 tornado, the most intense possible, then commissioned another risk assessment, which rated the likelihood of an accidental pathogen escape at 0.1 percent. When it opens in 2023, NBAF—like the new National Quarantine Unit for humans, in Omaha, Nebraska—will replace a complimentary cordon provided by the ocean with extraordinary engineering controls.

From our view onto the enormous construction site, this mostly seemed to take the form of an awful lot of concrete. "Enough to build a sidewalk from here to Oklahoma City," said Ron Trewyn. "Sixty thousand cubic yards, I believe, poured over two and a half years." The site itself was off-limits, but, over coffee in a nearby hotel lobby, the architect

Eugene Cole, who came to NBAF after leading the design of the CDC's new emerging infectious disease lab, gave us a tour of what lies beyond the concrete—itself a high-performance variety, with a built-in controlled chemical reaction that causes the concrete to expand after setting, leaving no room for cracks.

Cole, a soft-spoken southerner whose passion for animal welfare initially led him to veterinary school, until he realized that he didn't want to smell like formaldehyde for the rest of his life, is something of a star in the small world of biocontainment design. His work on Building 18 at the CDC received several awards and a special mention in *R&D Magazine*'s Lab of the Year feature, while NBAF, much to his delight, will achieve LEED certification, something many in the field considered impossible. Daylight and spaces for socializing are almost as important to Cole as the technical specifications. "How do I make the space appealing to the best researchers, when it's more or less a tornado bunker?" he said, describing the challenge. "You won't find many BSL-4 environments that have an outside window."

That said, there is no doubt that containment is NBAF's most important function. Cole told us that the windows he's managed to smuggle in are blast- and impact-resistant, with a metal grille on the outside to meet Nuclear Regulatory Commission guidelines for high-wind events. "But, in a tornado, it's the pressures that are difficult," he said. BSL-4 rooms are built using the "box-inside-a-box" principle, in which a negative-pressure lab is surrounded by a positive-pressure buffer, ensuring that air is always sucked inward, deeper into the building, rather than escaping out into the atmosphere.

The negative-pressure vortex that can form at the center of a slow-moving storm would seem to present a challenge to this system, but Cole assured us that NBAF will have a barometric reference loop buried deep in its mechanical core so that it's not thrown off by a sudden drop in external air pressure. Just to be sure, Cole also ran what he described as "containment-integrity testing"—manufacturing a closed negative-pressure bubble to see whether concrete around the embedded ductwork and plumbing would crack and leak under storm conditions. (It didn't.)

A BSL-4 laboratory, as he described it, is something like a layer cake: the laboratory floor where all the germs are being studied sits above an effluent-handling floor, and beneath a filtration level, a mechanical level, and an exhaust-venting "penthouse." All of those pipes and wires and ducts are in their own compartmentalized, containable space—but they also have to be accessible for regularly scheduled tests and preventative maintenance. This, Cole pointed out, will actually be the largest line item on NBAF's operating budget—simply operating a biocontainment facility costs significantly more than is ever spent on research. "Many times, from a design perspective, the focus is all on the science," he said. "That's a huge mistake." Cole and his colleagues have carefully designed pathways through the spaghetti of piping to make access as quick and easy as possible, and integrated a computerized maintenance management system so that the building all but tells its operating staff what it needs.

Pathways on the lab floor are equally carefully engineered so that people, animals, and stuff can move in only one direction: from "cold" or clean to "hot" or contaminated, exiting through a fumigation vestibule, a chemical dunk tank, an autoclave, or, in the case of humans, two chemical showers, and a personal shower, each in its own air lock. (One of the common complaints about Plum Island was that researchers were essentially required to shower communally, with only a curtain separating them from the adjacent corridor. "Times have changed," said Cole.)

Most aspects of a BSL-4 lab are the same no matter where you are, although Cole said that the U.K., as we saw in its human isolation facilities at the Royal Free Hospital, has a different culture of containment. Instead of the pressurized moon suits that researchers in the rest of the world wear to work with high-level pathogens, researchers at Porton Down (Britain's top-secret Defence Science and Technology Laboratory) use biosafety cabinets with built-in gloves to contain small animals, such as mice, while larger animals like pigs are held in a Trexler isolation unit. "It's a big bio-bubble," Cole, who is not a fan, said, where the researcher has to "crawl under and then pop up like the Michelin Man. It's odd."

Apart from being much bigger, NBAF does include some upgrades on its neighbor, the Biosecurity Research Institute. One area of improvement is its carcass-disposal methodology. "Technology keeps evolving around this stuff," said Cole. The older facility has a tissue digester that dissolves animals in an alkaline soup, leaving only "bone shadows"—the calcium phosphate outlines of bones and teeth, stripped of all organic material. The latter is dried and incinerated; the former—a soapy solution of amino acids and peptides—is sterile enough to be released into the municipal sewer system. The only issue is that it is still so full of organic molecules that it can easily overwhelm the capacity of the wastewater treatment plant, so, before every release, Cole explained, the BRI team has to call the city "to see if they're ready to take that slug." This usually occurs late at night: the corpses of liquefied animals passing through the town's sewers while Manhattan's residents sleep, blissfully unaware, in their homes above.

At NBAF, however, Cole said, "our carcass material will never go down the drain." Instead, the facility has two thermal tissue autoclaves—"basically a big pressure cooker with a paddle in it," he explained. The solution that results is sterile enough to safely use as fertilizer—but, out of an abundance of caution, it will be put into fifty-five-gallon drums and incinerated instead. "There are just redundancies on top of redundancies," Cole told us, describing the building's parallel filtration units, dual electric feeds, secondary substations, and backup generators.

The feature that Cole is probably most proud of is the flooring. Rather than use vinyl or tile, which could chip, peel, or crack, NBAF's floors are coated with a chemical that binds at a molecular level to concrete, forming a water-resistant layer that can be efficiently decontaminated again and again. To make sure that cows, sheep, and pigs won't slip, Cole knew he needed to mix in some grit—but not so much that it would tear up their hooves and cause discomfort.

No one had ever taken the time to scientifically determine what the right amount of grit should be, so Cole decided to do it himself, in his basement. He acquired a machine used to test the slip resistance of shoes on carpet and persuaded the necropsy lab at Kansas State veterinary school to donate some animal hooves. "They're like your fingernail, just big," he explained. Cole attached a series of these disembodied hooves to the machine's mechanical last—a human-foot shape—then set them in motion, stepping in place over and over again, while he measured the flooring material's friction and durability, as well as any hoof abrasion. "Yeah," he said, grinning, "my wife was not happy." Still, the outcome, in his opinion, is perfect flooring: cleanable, easy on the animal's feet, and completely nonslip. He has published his findings, and hopes to have them enshrined as a new international standard.

"We all suffer from OCD," Cole admitted with a slightly embarrassed laugh. "I mean, to be in containment design, you do have to be worried about the details."

Despite these efforts, the primary way that infectious material is expected to leave NBAF is due to human error—the Achilles heel of lazarettos throughout history. "Everyone will tell you it always comes down to the people," said Trewyn. NBAF will implement continuous training, voluminous record-keeping requirements, and a buddy system, but staff will not be allowed to keep chickens at home, on the off chance they bring home a pathogen that jumps species. (A disease that becomes avian is always of elevated concern, simply because of how easily it can move across borders.)

Everyone who works at NBAF will also be subject to background screening and security checks, to mitigate the possibility of an insider threat; the building is designed so that advancing deeper into containment involves passing through concentric rings of facial recognition and PIN-code checkpoints. Trewyn told us that, on the recommendation of a Navy SEAL white-hat security team, the university is already moving its Purebred Beef Unit, which offers students hands-on training in the cattle industry, to a new location, farther away from NBAF. "Somebody could poison the animals, once NBAF is up and running, and it would have nothing to do with NBAF, but that would be the perception," he explained.

In yet another report, the National Research Council complained that the Department of Homeland Security's risk assessment was "based on overly optimistic and unsupported estimates of human error rates." (The DHS assessment didn't even attempt to quantify the likelihood of malicious or deliberate acts.) Certainly, Plum Island has had a handful of documented close calls, as have other such facilities around the world. But Trewyn believes that the risk of accidental pathogen escape, in all its uncertainty, is worth taking, based on the probability that these diseases will arrive in the United States anyway and cause equally incalculable damage.

As an example, he pointed us, once again, to the U.K. In 2007, when foot-and-mouth escaped from the Pirbright Institute, an animal-disease research facility, into the Surrey countryside, thanks to a combination of heavy rains and aging pipework, it was quickly caught and contained. Within hours of the first case, all livestock movement in the entire country had been brought to a halt, and the virus was mopped up within two months, after infecting just eight farms. The system worked, Trewyn concluded, especially when compared with a very different incident just six years earlier.

That outbreak began in Northumberland, in 2001, when contaminated pork that had likely been illegally imported from Asia was fed to a herd of pigs in improperly sterilized swill, triggering a national epidemic of foot-and-mouth disease that lasted nearly a year and led to the death of six million sheep, pigs, and cattle, and at least sixty farmer suicides. Soldiers were brought in to assist with the mass slaughter of affected herds, and, following footage of the British countryside alight with animal pyres and bulldozers shoveling rigid carcasses into huge piles for incineration, tourism dropped by 10 percent.

Calculating the cost of such an outbreak is almost as tricky as assessing its risk: the slaughtered animals and their disposal is one thing, but there's also the lack of access to market for farmers with healthy animals, as well as restrictions on export and import. Animal diseases are unlikely to cause famine, but disruptions to the national meat supply can cause prices to skyrocket, leading to very unhappy consumers.

In 1902, when the price of kosher beef jumped from twelve to eighteen cents a pound, the women of New York's

Lower East Side rioted, breaking windows and throwing steaks. More recently, in the past couple of years, a quarter of the world's pigs have died from African swine fever, the deadliest outbreak of which most Americans have never heard. There is currently no vaccine; NBAF hopes to develop one, while other researchers are working on genetically modifying the swine genome to engineer resistance. In China, African swine fever has already claimed at least 40 percent of the country's pig population, and the price of pork has more than doubled—a serious problem for a commodity whose cost has roughly the same political significance in China as gasoline prices do in the United States.

As always during outbreaks, criminals have emerged to take advantage of the opportunity. According to a 2019 exposé by state news agency Xinhua, gangs have begun using drones to drop infected feed onto farms in areas that have yet to be affected by the disease. They then swoop in with an offer to buy the animals at a steep discount, in order to cull them, but instead—despite a national ban on pork and pig movement—they smuggle the herd back to affected provinces, in order to sell the meat where pork prices are even higher. The report claimed that one gang had smuggled as many as four thousand pigs between provinces in a single day, bribing inspectors and faking quarantine certificates to get the animals across checkpoints.

In response, one pig farmer in the country's northeast installed an anti-drone device that unfortunately also jammed the navigation systems of planes heading to a nearby airport. The country's largest pig producer has recently invested in twelve-story biosecure piggeries: each floor has its own air-handling and disinfection system to limit the spread of disease, while staff live in dedicated housing on-site, spending two days in quarantine every time they enter the facility, unable to leave the farm until their day off. One farmer in Hunan Province told a *New York Times* reporter that pigs have become so rare in his region that when he transports his animals for sale, people gather around the truck to stare. "It's like they were seeing a panda," he said.

Fifty countries have now confirmed the presence of African swine fever in their herds, as far afield as the Philippines and Poland. Denmark, a porcine powerhouse, has begun construction of a wild-boar-proof fence along the length of its border with Germany to keep the virus out. In Australia, sniffer dogs have been stationed at airports, and mail is screened in order to catch pork being smuggled into the country; the pathogen survives for months on surfaces and in even heavily processed and cooked meat. "Only one country has been able to eradicate this disease," the Australian agriculture minister told reporters, referring to the Czech Republic's successful four-year African swine fever elimination campaign. "They sent their army into forests night after night to shoot every single feral pig."

For many experts, the question is when, rather than if, African swine fever will arrive in the United States. Pig farmers in the United States have been advised to implement disinfection protocols at farm gates, ban foreign visitors, and inspect farmworkers' packed lunches for contraband bacon sandwiches or hot dogs. In 2013, more than 10 percent of American pigs died when a porcine epidemic diarrhea virus arrived in the United States on the reusable bulk bags used to transport feed; researchers at Kansas State University have shown that the half-life of African swine fever virus in shipped feed is two weeks.

"It has not come to North America—that's great," said General Myers, who, in 2016, moved to Manhattan to become president of Kansas State, his alma mater. "To say that it won't in this globalized economy of ours—that is probably a foolish statement for somebody to make."

Since Myers's troops discovered al-Qaeda's list of pathogens in an Afghan cave, there has been a huge investment in research, to which NBAF is testament, but not much in the way of local planning. Ron Trewyn told us of one local sheriff—the exception, rather than the rule—who had mapped the optimal locations for the forty roadblocks needed to create a cordon sanitaire, quarantining his entire county in the event of an outbreak; inside the exposure zone, according to the plan, "all cloven-hoofed animals would be destroyed."

"I wish I could say that every county in this state had that," said Trewyn. Myers agreed: "I think we're intellectually better prepared, but I don't know if we're operationally better prepared. Are we really ready to destroy millions of pigs?"

The logistics of what animal-health experts euphemistically term "depopulation activities" can rapidly become overwhelming. Faced with the need to cull almost eleven million pigs during a swine fever outbreak in the Netherlands, the government resorted to mobile electrocution devices, described chillingly by the journalist Maryn McKenna as "a pig-sized box that forced the animals to walk over a wet metal plate while zapping an electric current through their heads." In 2015, when avian flu meant that thirty-eight million chickens, ducks, and turkeys had to be slaughtered in Iowa, birds rotted on farms when local landfills refused to accept any more diseased carcasses, for fear of lawsuits from neighboring farms. "I've been in the landfill business probably twenty-six years, and I've never ever seen this kind of volume, and I hope I never do again," the director of the Northwest Iowa Area Solid Waste Agency told a local public radio affiliate.

Farmers are entitled to compensation for the animals they have sacrificed, although not for lost income due to restrictions in movement and other quarantine measures, or the wasted production time. What's less clear is who pays for it: the division between state and federal responsibilities is blurred. "How do you interdict the transportation network to make sure sick animals aren't moving around the United States infecting more herds," asked Myers, "and what authority do we have to stop them?"

Myers pointed out that, in case of emergency, the Department of Defense is responsible for providing enforcement for the USDA (and other federal agencies). "Local authorities generally get overwhelmed pretty quickly and then call the DOD," he said. But that is usually where the plan ends. "We actually practiced scenarios where you have people representing all the major departments and agencies of the U.S. government," Myers told us. "When we got to the point where it said, now we're going to call the DOD for help, the scenario would end."

"It's ludicrous, right?" he continued. "The scenario would end and we, the DOD, never got to play out what it is that was needed—is it communications, is it security, is it helicopters? What is it?"

⟷

"Whatcha got?" yelled Mitch Vega, over the roar of dozens of eighteen-wheelers. "Auto parts," the truck driver shouted back, handing Vega a clipboard of paperwork. "Yeah, you do!" Vega replied, returning it after a quick glance. "Have a good day!" He was wearing a high-visibility vest over his khaki uniform; a California Department of Food and Agriculture baseball cap, aviator sunglasses, and a silver walrus mustache completed the look. "How you doing, bud?

Good?" he yelled to the next driver in line. "Sausage casing," Vega told us, sotto voce. "We don't worry about that." He waved the truck driver on his way with a cheery, high-volume "Have a good day, bud!"

Vega told us he'd grown up nearby, in Needles, California. "I moved to Wisconsin, and, well, it didn't work out," he told us. "I love it here—every day is different." The only downside was bee season, which, when we visited the Needles Border Protection Station on a sunny, blue day in early February, was at its peak. Bee season, the otherwise imperturbable Vega confessed, "is horrible. It's four months, starting in November, and it's nonstop." What's more, he is allergic to bee stings, as was our host for the day, the station manager Michele Jacobsen.

California's agriculture industry is worth $50 billion: two-thirds of the United States' fruits and nuts are grown

An agricultural quarantine and inspection station on the California-Arizona border, as photographed by Dorothea Lange in 1937. (*Courtesy of the U.S. Library of Congress*)

there, a bounty that includes 100 percent of America's almonds. (California is responsible for an astonishing 81 percent of the world's almond supply.) All those almond trees flower in February, when they are visited by about seventy-five billion bees in what is considered the world's largest pollination event.

A sizable percentage of those bees come through the California Border Protection Station on I-40, a few miles south of Needles, near the Arizona state line. This four-lane checkpoint, complete with six staff members, two dumpsters, an incinerator, and a single-story office building with a Coke machine by the door, is one of sixteen such stations, all charged with quarantining California from the rest of the nation's agricultural pests and diseases.

"There you go, sir!" yelled Vega. "Mars Chocolate," he said to us, while waving another semi onward, its hubcap spinners glinting in the sunlight. "You figure, everything that we need, it comes through here." Vega shrugged. "There's some strange stuff, like autopsy table equipment."

Every fifteen minutes the inspectors rotate, at which point Vega escorted us back to a low-slung building that combined the functions of an office, laboratory, lunchroom, and bee-suit enrobing facility into a couple of overstuffed rooms. Cans of bug spray sat on the windowsill, and stacks of forms were wedged on the countertop next to a microscope, a cutting board and knife, half a kiwi, and a three-ring binder filled with printouts of quarantine orders for different pests, each in its own plastic sheet protector.

Kiwis are one of the few fruits and vegetables that are not subject to a quarantine order, but a random inspection earlier in the day had turned up some problematic bugs in a container from Italy, and the truck had been turned around. "What's *not* under quarantine is easier to tell you than what's under quarantine," said Jacobsen, an upbeat, pragmatic woman in her fifties. "A lot of your lettuces and your tomatoes are not under quarantine, your melons are not really under quarantine, and there's no quarantine on bananas or onions," she said. "Other than that?" She paused, thought for a moment, and then shook her head. "There's a lot of stuff under quarantine."

Jacobsen and her tiny team inspect every single truckload of produce and livestock—a category that includes bees—that is subject to a quarantine order, and then conduct random sampling on the rest, all the while trying to keep traffic flowing smoothly. Every other word of our conversation was punctuated by an email ping, a phone call, or the buzz of walkie-talkie transmissions between inspectors, but Jacobsen didn't seem frazzled, even when she realized that she'd received the wrong kind of toner for the station's only printer, which was now out of action on one of the busiest days of the year. "I was a single mother raising kids and working in a flower shop, and, on the weekends, I was helping a lady who ran a pest-control business," she told us. "I had college, but I didn't have enough, but, with my experience, it qualified me to get on as a seasonal, and get a shot at it from there."

Nearly two decades later, Jacobsen has been running the station for years and is visibly thrilled each time her team prevents an agricultural pest from entering the state. In 2014, she told us, the team at Needles was the first of the California border stations to intercept the Asian citrus psyllid, a tiny insect that looks like a grain of brown rice but carries a disease that has all but destroyed the citrus industry in Florida, wiping out five of every seven growers. As soon as the pest was identified, the infested tree went straight into the incinerator behind the station, and California's $7 billion citrus industry dodged a bullet. "When we find something, and it's the first time we've found it, we get really excited about that," she said.

New pests are always being added to the quarantine list, and with each comes a new set of cargoes to inspect. In some cases, this is straightforward: avocado scale pest is, no surprise, found on avocados. In others, it requires some creativity, or thinking like a pest. Jacobsen told us that she originally came up with the idea to start inspecting portable moving containers, like PODS or ReloCubes, because she realized that while the container is being loaded in people's front yards, it might be sitting on a fire ant mound. "We had a think-outside-the-box day where we try to come up with new pest pathways," she explained. "So we're like, let's start checking those."

As it turned out, rather than fire ants, the PODS they inspected were hosting another quarantine pest: gypsy moth egg masses. Now California Border Protection staff regularly check PODS, as well as all sorts of other things that sit outside for extended periods—patio furniture, barbecue grills, even old cars.

⟷

The gypsy moth is one of America's oldest and most destructive plant pests, following its deliberate importation by a French immigrant, Étienne Léopold Trouvelot, in 1869. Trouvelot was hoping to interbreed the European gypsy moth with Chinese silkworms to create a hardier American hybrid that could launch a national silk industry. He kept netting over his backyard hatchery, but the gypsy moths soon escaped, and, by the 1880s, his hometown of Medford, Massachusetts, was utterly infested. One neighbor, J. P. Dill, told *The Boston Post* that "there was not a place on the outside of the house where you could put your hand without touching caterpillars."

Before long, every single tree in the area had been entirely denuded of its leaves, branches stripped as bare as winter even at the height of summer. "We could plainly hear the noise of their nibbling at night, when all was still," said Dill. "It sounded like pattering of very fine raindrops. If we walked under the trees we got nothing less than a shower bath of caterpillars."

Trouvelot moved on to become an astronomer at Harvard; he has craters on Mars and the Moon named in his honor. Meanwhile, the gypsy moth has continued to spread across America at a rate of approximately thirteen miles every year for the past 150 years. Cutting down all trees in a protective cordon around the infestation, in addition to liberal applications of DDT and other pesticides, has failed to stop it from colonizing as far as Wisconsin to the west and Virginia to the south. The USDA estimates that, as of 2014, gypsy moths had single-handedly defoliated an area of hardwood forest at least four-fifths the size of California—but, thanks in part to Michele Jacobsen and her team, the moth is not yet established in that state.

The story of the gypsy moth neatly illustrates the warm welcome Americans initially extended to plants and animals—the 1800s were the heyday of acclimatization societies and plant explorers, dedicated to importing flora and fauna from the rest of the globe to enrich the American landscape—and the unpleasant reckoning that followed, when several of these new arrivals proved to be more destructive than useful.

California was the first state to try to curb the invasion: in 1881, following the discovery of cottony-cushion scale and red scale bugs on nursery stock from Australia and the pernicious San Jose scale bugs on trees from China, it set up a plant-inspection system. "We're naturally protected by boundaries—our mountain ranges, our deserts—and a lot of the agricultural pests are not native," Jacobsen said, explaining the state's fervor to protect its burgeoning orchards and vineyards. A few years earlier, a handful of European nations had implemented quarantine restrictions after an estimated 90 percent of the continent's own vineyards were lost to an American aphid, phylloxera. Just like cholera, this lethal pest proved capable of surviving intercontinental voyages once the invention of steamships compressed the natural quarantine afforded by the ocean crossing.

The federal government took a little more persuading, even after the gypsy moth debacle. First came chestnut blight, brought in on imported Japanese chestnuts in the early 1900s. The American chestnut has been described by arboriculturists as the perfect tree: it was tall and fast-growing, with rot-resistant, straight-grained wood that was perfect for log cabins, telephone poles, and railway ties, and it yielded an abundant harvest of delicious nuts. By 1940, all four billion of North America's chestnut trees had been wiped out.

The final straw, however, was the cherry tree incident of 1911: the photogenic, pink-blossomed trees that surround the Tidal Basin in Washington, D.C., are actually replacements for the original gift of two thousand cherry trees from the people of Tokyo, which were burned in great piles on the Washington Monument grounds after USDA officials realized they were infested with two different plant pests, one plant disease, and an unknown species of moth. The embarrassment created by this diplomatic faux pas finally nudged Congress into action, and the first federal Plant Quarantine Act was passed the very next year.

⟷

Back at the Needles station, the bulk of the work consists of inspecting and sampling commercial loads of fruits, vegetables, and, of course, bees. "We're profiling vehicles, too," said Jacobsen. "If they're coming from far away, we ask to look in ice chests." If an avocado or orange is lurking in there, Jacobsen and her team confiscate it.

"If it's a RV, we ask to look in their shower, too, because, a lot of times, that's where they keep their little house plants that they sit outside when they're camping," she said. Increasingly, Jacobsen said, wooden sculptures have started turning up during random inspections, thanks to the growing popularity of chainsaw carving. "It's beautiful, but you start seeing insect boring-hole damage, and then, next thing you know, you're pulling something out with the tweezers," she said.

Jacobsen and her team were universally polite and pleasant—almost shockingly so in a country where officials too often seem to consider basic courtesy a sign of weakness, or, at least, an optional extra. Still, not everyone responds well to forfeiting their fruit and firewood at the border. Some try to smuggle their contraband through anyway: Jacobsen told us about a recent incident involving a bag of pecans, which can carry weevils and husk flies. Their owner refused to hand them over, and was thus refused entry; a few hours later, he showed up again with the pecans hidden in his suitcase. "He said he forgot they were in there," said Jacobsen, with a small smile.

Others are so irrationally anxious about being caught with anything subject to quarantine that the side of the freeway approaching the inspection station is littered with discarded watermelons, bananas, and baby carrots, all of which are perfectly admissible. "We get a lot of rude comments and stuff, but that's okay," said Jacobsen. "It's just a lack of education."

The public, in general, is blissfully ignorant of the dangers posed by plant pests and diseases, as well as the enormous efforts required to eradicate them. Take the Mediterranean fruit fly, or medfly, one of the pests that Jacobsen and her team look for during every inspection. Economists have estimated that, if it were to become established in California, the medfly could cost the state more than a billion dollars annually in fruit destroyed by larvae, as well as lost export business.

In an attempt to prevent its spread north from Mexico, the U.S. government helps fund the largest fruit-fly rearing facility in the world: a cluster of warehouses in Guatemala capable of producing several billion male pupae per week, which are then loaded into a cobalt-60 powered irradiator to be sterilized, before being dyed fluorescent pink or orange for identification purposes. The idea is that female fruit flies will mate with these sterile males, and effectively eradicate themselves.

Since 1996, billions of those irradiated flies have ended up in Los Angeles, loaded into small planes that dispense 32,500 insects per linear mile from chutes that stick out from the bottom of the fuselage. The flight paths of the planes are so low and so unusual—they cover the basin in long, parallel sweeps, as if going up and down every aisle in a supermarket—that, during the city's widespread Black Lives Matter protests in June 2020, some activists suspected the aircraft were conducting covert surveillance.

This preventative program costs $16 million a year, according to the California Department of Food and Agriculture, but it has reduced the number of new infestations by more than 90 percent. (The state also maintains ten fruit fly traps per square mile on the ground, checked at weekly intervals; when inspectors find a pink or orange fly, they know not to worry about a new invasion.)

Chasteningly, given their expense, these surveillance and control measures are ultimately expected to fail. Infected and infested material inevitably slips through. At Needles, Jacobsen and her team can hardly keep up with the flow of traffic; monitoring the entire U.S. border is unfeasible. Currently, only 2 percent of containers are inspected at the Port of New York and New Jersey; increasing that to just 5 percent would cost an extra $1.2 million per month.

Even more explicitly than those for their human counterparts, plant quarantine measures are subject to a sort of decay, decreasing in predictable half-lives of efficacy and expedience until they are abandoned altogether. The cost of inspection, surveillance, prevention, and other plant quarantine measures is simply the purchase price of time in which to develop treatments—pesticides and fungicides—and, hopefully, breed resistance.

The bees passing through Needles are an extreme example of quarantine decay in action. While we talked with

Inspecting bees at the agricultural quarantine and inspection station outside Needles, California, near the border with Arizona. (*Photograph by Nicola Twilley*)

Jacobsen in the crowded office, another inspector was sending a microscope photograph of a bug he'd found on a load of bees over to the California Department of Food and Agriculture's diagnostics unit in Sacramento. He'd found it between the beehive boxes, where fire ants often nest; carpenter ants and termites are also common in any rotten wood. The team in Sacramento would, when they got back from lunch, identify the bug, and, more important, determine whether it was "Q-rated"—a destructive insect that isn't yet established in the state.

Bee suits on, we went to visit the trailer full of hives parked in the pullout, awaiting the verdict. Next to it were bees trucked from Florida, Louisiana, Texas, and South Dakota, their buzz still audible above the roar of traffic. All had been rejected for quarantine-rated pests; several were being reloaded after being pressure-cleaned, and would line up for reinspection later in the day. "Our thing is we don't want to stop them from coming in," said Jacobsen. "We just want them to come in clean."

"Clean," for bees coming into California, used to mean free of *Varroa destructor*—a tiny, two-millimeter-long Asian parasitic mite that arrived in the United States more than thirty years ago and is blamed for much of the terrifying collapse in honeybee populations over the past two decades. In just one winter, from fall 2018 to spring 2019, *Varroa* is estimated to have killed an incredible fifty billion bees, a figure equivalent to more than seven times the entire global population of humans.

Back in 1988, more than a decade before Jacobsen began working at the border, California was free of the pest, and bees from Florida and South Dakota were under quarantine for it. The first recorded case in the United States had been found a year earlier, in some hives transported from Florida to Wisconsin. By 1995, every state in America had been affected; and by 2005, California had entirely given up trying to restrict the mite's entry, focusing on fire ants instead.

Today, beekeepers regard California's almond pollination season as a super-spreader event, and *Varroa* mites are beginning to develop resistance to the leading pesticides. Fortunately, researchers say they are close to breeding a so-called hygienic bee, capable of detecting and removing mites from the hive on its own. Quarantine simply bought a little time, before its leaks and its costs decayed its utility into obsolescence.

⟷

There is a corner of England that is forever West African: the International Cocoa Quarantine Centre in suburban Reading, just outside London. This recently constructed greenhouse, a $1.5 million purpose-built facility nearly the size of an Olympic swimming pool, is the world's cacao buffer zone, shielding humanity from an otherwise chocolate-free future. It is also a rare example of true quarantine in the universe of nonhuman biosecurity.

In most cases, if animals and vegetables are suspected of having been exposed to an infectious disease, they are destroyed rather than quarantined. The actuarial calculus—the risk posed to their destination weighed against the value of their arrival—does not come out in their favor. Cattle, pigs, and chickens are culled en masse; plants, seeds, and vegetable matter are incinerated, or, if possible, cleaned. Biosecurity is still important, as we had discovered—it just tends to consist of total containment, as at the new National Bio and Agro-Defense Facility in Manhattan, Kansas, or surveillance and border control, as at Needles.

For the most part, only companion and sporting animals—pets and horses, plus the occasional panda en route to a foreign zoo—are considered worthy of the time and space required to perform true quarantine. Even then, most countries no longer detain microchipped pets with up-to-date health passports; when the ARK, a new, $65 million luxury quarantine facility for animals, opened at JFK in 2017, it sat largely unused. By April 2020, only a last-ditch pivot to handling medical eggs for vaccine production was keeping it afloat. Australia is a singular holdout: its quarantine requirements are so stringent that equestrian events at the Melbourne Olympics were actually held in Sweden, and Johnny Depp's smuggled Yorkshire terriers faced deportation or death. (This rigorous enforcement has led to occasional disaster, as in 2017, when the National Museum of Natural History in Paris loaned a collection of rare herbarium sheets and plant-type specimens, some dating back to the 1700s, to the Queensland Herbarium: they were promptly incinerated by Australian authorities in what *The New York Times* called a "quarantine flub.")

As with that for animals, "quarantine" for plants typically refers to import bans, inspection, or fumigation. Most plant material is simply not valuable or unique enough to warrant a dedicated facility and the time to grow and screen a seedling to make sure it poses no risk.

The exceptions are few. Kew Gardens, in southwest London, has a state-of-the-art quarantine greenhouse that acts as a "plant reception," ensuring that new arrivals do not jeopardize the largest and most diverse botanical collection in the world. The tree that the French president Emmanuel Macron planted with Donald Trump on the White House lawn in 2018—an oak sapling from Belleau Wood, site of intense fighting in World War I, meant to symbolize the two countries' historic friendship—was dug up the next day and put into quarantine at the USDA's facilities in Beltsville, Maryland. Its roots had reportedly been "enclosed in a special plastic coating" for the ceremonial event, so as not to risk contaminating American soil. The plan was to release and replant the tree if, at the end of two years' monitoring, it proved pathogen-free, but a diplomatic source told Agence France-Presse that, in 2019, amid reports of worsening relations between the two leaders, it had died.

And, of course, there is the International Cocoa Quarantine Centre, a few miles west of London, in the Berkshire commuter belt. Stepping inside its polytunnel—a high-tech hoop greenhouse covered in polyethylene—offered a delightful contrast to the brisk March day of our visit. Roughly four hundred different varieties of cacao plant were growing in a computer-controlled environment that mimics their tropical rain forest home.

Chocolate sits in quarantine's sweet spot, where the scale of the threat and the value of the market make housing cacao plants for two years in a purpose-built greenhouse on some of the United Kingdom's more expensive real estate seem worthwhile. While chocolate demand keeps growing, with consumption rates in new markets such as China and India doubling year on year, the supply is shrinking. Social and political shifts, including increased urbanization in West Africa, which grows more than 70 percent of the world's supply, are part of the problem, but the real challenge comes from the pest- and disease-prone nature of the cacao plant itself. Among plants, cacao is unusually unlucky—or attractive—in that it seems to gather new pests and diseases wherever it is grown, while remaining susceptible to the old ones. A 2010 trade guide to the international cacao industry concluded that "the sheer number of known diseases and pests that attack cocoa makes one wonder how enough chocolate bars are ever produced."

The result, according to gloomy headlines, is a looming "chocpocalypse." In little more than a decade, according to a Ghanaian nonprofit, a Hershey bar may well be as rare and expensive as caviar.

Today, cacao is still grown in its South and Central American center of origin, but the majority of the world's supply comes from West Africa, and, increasingly, Southeast Asia. "Each of these three areas have got their own pests and diseases," explained Paul Hadley, a tweedy and bespectacled cacao researcher at the University of Reading, reeling off a list of unpleasant-sounding names: cocoa pod borer, frosty pod, vascular streak dieback, cacao swollen shoot virus, witches' broom. "Those pests and diseases are reducing potential cacao yields by about thirty percent already," he said. "So you really don't want to make it even worse by transferring any of them from one cacao-growing region to another." Indeed, if frosty pod and witches' broom made the jump from South America, where they have already devastated cacao production, to West Africa, "the impact would be appalling," according to Hadley. "That would be curtains for chocolate."

Avoiding that bitter fate is the primary goal of the International Cocoa Quarantine Centre, which is funded by a consortium that includes the USDA and the London Cocoa Futures Market as well as major confectionary companies such as Mars. Growers and researchers in Côte d'Ivoire, Ghana, and Indonesia want access to different varieties stored by the International Cocoa Genebank in Trinidad, as well as to promising new varieties generated in breeding programs in Costa Rica, Ecuador, or Brazil—but specimens from those countries could well be invisible carriers of disease. Frosty pod can lurk in a plant, asymptomatically, for months. During this state of limbo—not recognizably sick, but not yet proven safe—a cacao plant en route for, say, Ghana, from its origin in Ecuador will simply lay over in sunny Reading for a few years. In fact, Hadley told us that more than 95 percent of the world movement of cacao genetic material comes through his greenhouse.

In the case of cacao, plants usually travel as budwood: a short length of branch with a bud that is then grafted onto a seedling, from where it grows into a genetic clone of its mother plant. Inside the Reading greenhouse, the technician Heather Lake showed us a shipment that had arrived from South America just a few days earlier: after grafting three buds onto separate seedlings, each in its own pot, she had put the new arrivals into insect-proof cages—individual white cubes where they would stay for the next couple of months, just in case there were any larvae on the budwood.

After they are given the all-clear for insects, the seedlings grow for another nine months until they are mature enough to produce buds of their own. Those buds are then grafted onto another rootstock, from an "indicator" varietal that has a useful tendency to show clear symptoms of any viral infections. Those test plants, arranged in a row in the center of the hoop greenhouse, are checked weekly for any signs of disease, while their mothers are lined up around the edge of the tunnel like anxious onlookers. If, after two years, they are deemed healthy, the indicator plants can be destroyed, and the mother plant's budwood is deemed safe to be harvested and sent onward to its destination.

"This is a slow process," Hadley admitted. "If you ask me for a particular variety, it will take a minimum of three years before we actually deliver something." New molecular tests for the viral diseases offer the potential for speeding things up slightly, but plant pathologists prefer to avoid risk wherever possible—"the belt and braces approach," as Lake put it.

Meanwhile, everything is done to keep the plants safe and happy during their stay: netting on the air-intake fans keeps British insects out, automated thermal screens and sunshades keep the heat in and the sun out, to mimic rain forest understory conditions, and a high-tech hydroponic system delivers drip feeds on a two-hourly cycle. The main threats, according to Hadley, come from vandals or power failures: if the temperature drops to fifty-three degrees or below for more than a minute or two, the plants will die. Before these new polytunnels were built, quarantined cacao was housed in a collection of older buildings located much closer to a main road, and people occasionally broke in, looking for something to steal. The new facility is walled off, which not only helps Hadley sleep at night but also acts as a layer of insulation, lowering the otherwise exorbitant heating bill a little.

Although the energy required to transform Reading's typical mild drizzle into the humid heat that suits cacao is not insignificant, it is precisely that gap between tropical and temperate conditions that makes England an ideal quarantine zone for cacao. There are no cacao plantations in the U.K. to be threatened with infection, and, in any case, any harmful pests or diseases on arriving cacao plants are adapted to rain forest conditions and are thus unlikely to stage a successful escape into the Berkshire countryside. For a certain class of high-value tropical crop, these "intermediate" periods of climatic quarantine in Europe or parts of the United States are quite common: banana plants are hosted by Belgium, coffee bushes by Portugal and Florida, and rubber trees in USDA greenhouses in Maryland.

"It is rather nice working here in the winter," Hadley said, escorting us back out into the damp gloom, the greenhouses protecting the world's chocolate glowing behind him like a beacon.

"All that bare earth and clay," said Roger Buckley, a British civil servant. "It's only the grasses that have gone, of course, but it's surprising to realize what a large amount of territory is covered with grasses of one kind or another."

The Chung-Li virus initially affected only rice, but the failure of this staple led to a famine that caused more than two hundred million deaths in China alone. Widespread application of a chemical treatment developed by Chinese scientists to control the disease instead selected for a more virulent strain: the new-and-improved Chung-Li extended its range to encompass all grasses. Wheat, rye, oats, and barley—even the turfgrass on which beef and dairy cattle depend—all died, as wind bore the virus's spores relentlessly onward, across Asia to Europe. Barricades went up to stop city dwellers from fleeing to the countryside and stripping the fields of what food remained, and the British government fell as it became clear that aid ships from Australia and America were never going to arrive.

This, of course, is the plot of *The Death of Grass*, a 1957 novel written by Sam Youd under the pen name John Christopher. But, for scientists, it's an only slightly exaggerated description of what might happen if a staple crop succumbs to the plant world's version of Disease X. After all, just fifteen crops provide 90 percent of the world's food. As the Kansas State University plant pathologist Jim Stack has been known to say: we're only one pathogen away from starvation.

"Could it happen? Yes. Is it likely to happen? No," added Richard Myers, the retired general who is now president of Kansas State University. "But you get a pathogen, let's say, in a wheat crop, and it spreads around the world—that's the sort of thing Jim's talking about."

At the USDA's Cereal Disease Lab in St. Paul, Minnesota, researchers work with that kind of pathogen—but

The USDA's Cereal Disease Laboratory in St. Paul, Minnesota. (*Photograph by Nicola Twilley*)

only from December through the end of February, in order to add a supplemental layer of thermal containment to the technical barriers built into this biosecure facility. With nothing green on the ground, and the landscape isolated under a blanket of winter snow, a rogue spore that escapes will simply die.

To get inside the double-tempered, double-paned greenhouse ourselves, we had to strip off: the pre-visit online training module warned prospective visitors to "enter the Shower Room naked with nothing in your hands unless you plan to not ever bring it back out again." We followed instructions in separate his and hers shower cubicles, and emerged clad only in blue Tyvek onesies, socks, and color-coded Crocs (pink for female visitors, blue for men).

Stephanie Dahl, the lab's quarantine officer, told us that researchers who were uncomfortable going commando under the Tyvek could bring in their own "unmentionables," on the understanding that the only way to get them out again was through the autoclave. The Crocs represented a major leap forward in biosecurity, Dahl explained—their foam resin surfaces were much easier to decontaminate than the canvas shoes they replaced. Unfortunately, there weren't quite enough for every researcher to have their own pair, but the addition of socks made it possible to share. "People are going to cringe when they hear that one," she said. "This facility is so phenomenal, and it does such great things, but people are really hung up on having to remove all your clothing."

Infected wheat plants at the Cereal Disease Laboratory in Minnesota. (*Photograph by Nicola Twilley*)

Inside the lab, a constant low-level whoosh was testament to the powerful air-handling system. Strips of plastic flapped under every vent as a visual confirmation that spore-laden air was being sucked in the right direction, toward the center of the facility, where it would be filtered before release. Wheat plants were lined up in dew chambers or under misting nozzles, as well as stacked haphazardly onto trolleys headed to the incinerator. On a metal tray, dozens of tiny, sickly specimens were growing inside individual cellophane bags, like gift-wrapped chives; a closer look revealed that several stems bore the telltale yellow-orange pustules that confirmed they were infected. Farther down the row, a technician was unwrapping their cousins one by one, each in its own plexiglass cube. He used a delicate little tool, handmade in Texas, to scrape at the lesions and funnel the resulting rusty dust into a pill-size plastic capsule in one smooth movement.

This was stem rust, a fungus that can regularly reduce a region's grain harvest by half or more, turning fields of golden-green wheat into a mass of blackened, broken stems and shriveled grains. As befits the nemesis of one of our oldest domesticated crops, rust has left its mark on human history. Its spores have been found at three-thousand-year-old archaeological sites in the Middle East. In ancient Rome, April 25 marked the annual festival of Robigalia, in which rust-colored animals—foxes, dogs, and even cows—were sacrificed in an appeal to Robigo, the goddess of rust. Ovid transcribes the priest's prayer in *Fasti*, a book of poems from 8 CE: "Scaly Robigo, god of rust, spare Ceres' grain; let silky blades quiver on the soil's skin . . . Spare us, I pray, keep your scabrous hands from the harvest. Harm no crops. The power to harm is enough." Robigo seems to have been unmoved by these pleas: historical weather records suggest that a series of rust-reduced harvests likely contributed to the fall of the Roman Empire.

The very first plant protection laws on record concern rust: in 1805, the principality of Schaumburg-Lippe, in present-day Germany, ordered the elimination of all barberry bushes; a German source mentions an even earlier antibarberry law passed in the French city of Rouen in 1660. The barberry, a sturdy, thorny shrub whose bright pink berries are deliciously sour, plays host to rust after the wheat harvest and before the new season's crop is sown. It was introduced deliberately into the United States for use in hedgerows, but, when a rust outbreak in 1916 destroyed about 40 percent of the nation's wheat crop just prior to the nation's entrance into World War I, the USDA decided barberry's time was up. The University of Minnesota, where the Cereal Disease Lab is housed today, helped launch a comprehensive barberry-eradication campaign, recruiting schoolchildren and Boy Scouts to distribute flyers and help track down barberries in backyards and parks, and hiring teams of laborers to dig up the bushes and then salt the ground in which they'd grown, to prevent regrowth.

Thanks to the successful elimination of the barberry in America's breadbasket states, and, even more significantly, an enormous investment on the part of the Rockefeller Foundation and the Mexican government to breed rust-resistant varieties in the 1950s, by the second half of the twentieth century, rust was no longer a concern. The Mexican program's leader, Norman Borlaug, was awarded the Nobel Peace Prize for his efforts, and untold famines were averted.

Today, however, said Yue Jin, a plant pathologist at the Cereal Disease Lab, "everywhere you look, the common barberry is coming back." Meanwhile, Jin's colleague Les Szabo continued, a resistance gene typically lasts only five years in the field before the rust mutates sufficiently to render it useless. In the 1950s, Borlaug had stacked a handful of different resistance genes in his Mexican varieties to try to create durable resistance. "The problem is the ball got dropped," Szabo said. "They kept just saying, OK, well, we lost that one, but we still have these—until the last gene went down."

The modern monocultures of industrial agriculture, in which the same handful of varieties of wheat are grown in every field, form an evolutionary forcing funnel, Szabo explained, inadvertently but aggressively selecting for mutations in rust that allow it to circumvent wheat's genetic armor. Over the past two decades, several such strains have emerged. From an initial mutation detected in Uganda and described in 1999, a deadly rust—christened Ug99—spread to Kenya. Farmers in the wheat-growing region of Narok, on the Ugandan border, lost 80 percent of their crop. From there, it has continued to spread and mutate, devastating harvests as far afield as Sicily and Iran.

At the Cereal Disease Lab, which is one of the few labs allowed to work with Ug99 in North America, the assumption is that this new super-rust and its descendants will eventually arrive in the United States—most likely via a stray spore on the hem of an international traveler's trousers. (This, they told us, is how rust first made it to the virgin territory of Australia in the 1970s.)

The ease with which rust spores cling to clothing and paper is the reason why neither ever exits the Cereal Disease Lab. The notes we made while talking with researchers were scanned and emailed to us, and the hard copy destroyed. We discarded our Tyvek and Crocs inside the hot zone, to be cleaned and reused, then stepped through an air lock into the shower for a full-body wash, before exiting on the other side to change back into our regular clothes.

These elaborate precautions keep rust contained in the lab, but, once it is in the field, there is no quarantine measure that can stop or even slow its movement. In the 1940s and '50s, researchers mounted slides coated in Vaseline on wooden sticks and held them out of airplane windows to map the "*Puccinia* pathway"—*Puccinia graminis* is rust's scientific name—an aerial superhighway that pulls spores from Mexico through Texas to Kansas all the way up to Canada. "A single pustule can produce ten thousand spores per day," said Szabo. "Now extrapolate that to a wheat field where each plant could have hundreds of pustules, and you can quickly see: once you find it in the field, the cat's out of the bag."

"We take quarantine measures when we work with or move the rust cultures," added Jin. "We do the screening in containment, in the winter, under very restrictive quarantine measures—but the disease itself is not a quarantinable disease."

Thus, while they work on developing resistant varieties, the only other thing to do is watch and wait. To provide an early warning system, Jin and his colleagues maintain "sentinel plots" along the southern border of the United States, in Texas and Arizona, as well as dotted around the Plains states. These, he told us, consist of just a few rows of a special indicator variety planted in a regular field. "They are watched over primarily by breeders," Jin explained. In the past, he said, the sentinel plot system was much larger, and farmers were paid to plant and monitor them, but the funding for that initiative dried up.

The use of these kinds of canaries in the coal mine of public health is expensive but effective. In Australia, which is the only continent free of the *Varroa* mite, sentinel beehives are stationed near ports, to alert biosecurity officials to any accidental introduction. In California, 139 flocks of sentinel chickens stand guard in chicken coops around the state; if the white leghorns are bitten by mosquitoes infected by West Nile or St. Louis encephalitis, they will develop antibodies that alert local public health agencies to the presence of the disease.

In the case of wheat rust, a sentinel plot can provide enough warning to sow a different crop, or at least to stock up on fungicides, to try to save some of the harvest. In the United States and Europe, farmers can usually afford the chemicals to treat their fields. (Of course, many of these are toxic to a broad range of organisms, and often run off into streams and rivers.) "In some countries, especially the poor farmers, they can't afford fungicides," said Jin. "Often, their wheat crop is their food crop *and* their cash crop, and rust will come through and wipe them out."

This is the real threat, clarified Jim Stack at Kansas State University. For now, the United States is still rich enough that there is likely no single plant pathogen that could reduce its population to starvation. But there are large parts of the world where a disease like Ug99 not only will destroy grain that could have fed millions of hungry people, but also will leave farmers penniless—without the cash to meet their own subsistence needs, let alone purchase seeds for next season.

"Hungry people are unhappy people," said Richard Myers—and, he warned, in today's interconnected world, famine and unrest are unlikely to remain local in their impact.

The other problem, Stack told us, is that while, at least in the United States, we may not actually be one pathogen away from starvation, we are also never faced with only one threatening pathogen at a time. The twenty-first century has seen an exponential increase in both the speed and volume of trade and travel—an increase that is redistributing animal and plant pests and pathogens, as well as human diseases, around the planet at an unprecedented rate. Collectively, Stack said, they could have the same impact as one superpathogen.

The number of alien species that have been imported into countries has increased dramatically over the past quarter century, an increase that correlates with the boost in global trade following the founding of the World Trade Organization in 1995. (WTO rules also legally require countries to use the least restrictive quarantine measures possible.) Researchers have found that you can calculate the rise in a European country's GDP based solely on increases in the detection rate of newly arrived, nonnative spiders.

Meanwhile, the containerization of global trade has introduced a new pathway: the crevices of these ubiquitous corrugated steel boxes offer a perfect berth for stowaway plant matter and pests to travel the world. At the Port of Brisbane, Australia, entomologists found more than one thousand live insects, many of them quarantinable pests, in a survey of three thousand containers—or little more than a day's worth of container imports for the country as a whole.

Most alien species are, of course, not only *not* a threat but, in fact, considered essential: wheat is both an alien grass in North America and the nation's staple grain. In the United States, there are seventeen thousand known native plant species and at least five thousand introduced ones, raising the question of whether it even makes sense to quarantine a landscape that is already one-quarter alien against further biological pollution. After all, the majority of these new arrivals will not become established, and many of those that do will cause no harm.

Unfortunately, as we'd seen, the handful that do pose a threat can wreak havoc. Agricultural diseases and pests lead to huge economic losses, as well as famine and food insecurity; some livestock pathogens also have the ability to jump species and infect humans. Nonnative plants, animals, microbes, and insects can also disrupt entire ecosystems—they can eradicate native species either by outcompeting them for food or by interbreeding with them, in what biologists call "gene flow." Those shifts, in turn, have ripple effects; they can transform a landscape's fire risk, pollination potential, and hydrological behavior.

Stopping trade for plant and animal diseases is not typically considered an acceptable response—even slowing it is met with resistance. As we removed our bee suits and prepared to leave the Needles station, Michele Jacobsen was sending Mitch Vega back out to open up the third lane, in order to keep traffic flowing. Despite the commitment and ingenuity of everyone we met, quarantine seemed less of a solution than a stopgap, its ambitions limited and its failure inevitable. But that, it turned out, was not necessarily the case.

⟷

"Chevron don't like to hear you say this, but Barrow Island is the gold-plated version of quarantine," the plant pathologist Simon McKirdy told us. Australia, by virtue of having been isolated as an island-continent for some thirty-five million years, is already home to an extraordinary biological diversity: more than 80 percent of its animals and plants are found nowhere else. Barrow Island separated from northwestern Australia eight thousand years ago, and is home to dozens of species that are now extinct or endangered on the mainland, as well as several species found nowhere else on Earth: a six-foot spotted lizard that runs as fast as Usain Bolt, a rat-size kangaroo, and a beaked, blind snake that looks like an oversize earthworm.

"It's probably the most pristine island around Australia, so it's the closest to what Australia used to be," said McKirdy. "It's one of the few islands in the world that all the common invasives—you know, the common house mouse, the brown rat—haven't been allowed to establish."

Barrow Island also sits above trillions of cubic feet of oil and natural gas, which the California-based energy company Chevron is keen to extract. The Australian government decided to allow the company to go forward with what has become one of the world's largest natural gas drilling platforms—the ominously named Gorgon Project—on the condition that it maintained the island's ecosystem intact.

This seems like a recipe for disaster, but McKirdy told us it was, in fact, the opposite: Chevron has the cash to do quarantine right. "Because they were spending seventy billion dollars building this plant anyway, they threw real resources to develop this," said McKirdy, who worked with Chevron to create a comprehensive Quarantine Management System that, he claimed, has thus far kept all nonindigenous species off the island.

The scope of the project is mind-boggling: there is no detail too small to be covered by a checklist, procedure, and guideline. "There's thirteen pathways for things to get to the island," said McKirdy, which included everything from people's luggage to helicopter transfers. "Every pathway has a process," he said—a layered series of interventions whose rigor and redundancy aim to intercept anything using that pathway. All fresh food headed to the island, for example, is washed, packaged, and inspected before it ships, and again as it arrives. Similarly, any mail is screened by human inspectors, detector dogs, and X-rays before being sent to the island.

"Everyone thought at the start that the biggest risk was containers," said McKirdy. "The standard container collects an amazing amount of crap. So, early on, they basically said, no, we just can't use that." Instead, Chevron manufactured its own fleet of shipping containers, with a sealed footing at the base, no vents, and a steel floor as opposed to the standard wooden one. The redesign didn't add significant expense or reduce capacity; there's no reason, McKirdy says, why it couldn't be rolled out globally, reducing the enormous number of stowaway pests that currently travel the world that way.

Once the containers had been dealt with, the leakiest pathway ended up being the workforce. "Just humans being humans," as McKirdy put it. Chevron hired its own inspectors to screen anyone traveling to the island for insects, fresh fruit, vegetables, seeds, and soil. To make sure those inspections were effective, McKirdy worked with a gaming company to develop *Quarantine Hero*, a biosecurity simulator that scored players on how many quarantine risk materials they found per passenger or item.

Flipping open his laptop, McKirdy launched *Quarantine Hero: Airport Module*, and checked his first passenger. Lucy was a smiling redhead in sunglasses and shorts who was, it turns out, carrying an illicit Snickers bar, and had a khaki weed thorn on her canvas shoes. At the end of the module, the screen said, "Well done, your shift is over," and showed McKirdy's final *Quarantine Hero* score, along with hot spots, quarantine risk material missed, and time taken.

"Inspectors competed against each other, and it just really took off," said McKirdy. He told us that Chinese quarantine authorities have partnered with the game's Australian designers to develop a version for their own inspectors.

Technology has also helped mop up the few invaders that have made it through the border inspections—the odd cockroach and dandelion. Clicking around on his desktop, McKirdy showed us a video of an autonomous robot, originally developed for use on farms, trundling over the island's reddish soil. "We're working with Chevron to adapt it," he said. "It's looking for weeds, and if it identifies one, this nozzle comes on and just sprays." McKirdy intends to tweak the robot so it can also create a georeferenced alert; we immediately imagined a fleet of rust-identifying, GPS-located robots roaming America's wheat fields in place of Yue Jin's more limited, expensive-to-maintain sentinel plots.

Similarly, Chevron's quarantine team has put seventy acoustic sensors all over the island, feeding into an AI system that listens for the particular chirp of the nonnative Asian house gecko, in case it somehow slips onto the island. "We're also working on using track pads that can identify the animal by footprints, tail prints, and weight," McKirdy added. "It's just trying to take some of those very slow, labor-intensive detection components out of the system."

Chevron, as one of the world's wealthiest corporations, is perhaps anomalous in the resources it can afford to devote to quarantine, but McKirdy thinks that its investment can pay dividends for all of us. Like Marty Cetron at the CDC, McKirdy wanted to know whether quarantine actually worked, and he has used Chevron funding to perform the detailed economic and scientific analyses needed to see which interventions are the most effective. "We've now got the statistical confidence that what we're doing is giving us the best result," he said. "Which, to me, is critical, and it was never really invested in before."

This, he hopes, could help governments, with their more limited budgets, make better choices about where to focus their time and money. "For example, governments would probably look at containers and just say there are so many, it's too hard to deal with," he said. "But the reality was, you need to devote effort at the container end and address that."

Another lesson involved something that Michele Jacobsen of the Needles inspection station had discovered by accident, thanks to PODS: thinking about pathways is more significant than targeting a specific pest. "With climate change, how do any of us predict what we're going to see?" McKirdy said. "We've got to have resilient systems, such that no matter what hits us, we can respond."

But, perhaps most important, McKirdy says that Barrow Island has demonstrated that quarantine *can* work—it can keep pests and diseases out while not impinging on the profitability of trade. "Oh, they've made it back in spades," he said, when we asked whether Chevron's extraordinary investment in biosecurity had turned the project into a loss leader whose only upside was an attempt to buy the company some environmental credibility.

Indeed, rather than the governments having to shoulder the burden of biosecurity on their own, Barrow Island has helped make the case that corporations can reasonably pick up some of the costs incurred by commerce and trade, as well as its benefits. "Really, though, it's about everyone being involved," said McKirdy. "If we all do our bit, we've got the best chance at getting the right outcome."

7

A Million Years of Isolation

Our descent into the Earth began noisily, the machine sounds of the elevator echoing back and forth within the concrete walls of its shaft. Unable to hear much, we didn't speak, until Bobby St. John reassured us with a shout: "We should hit salt soon!" St. John, an affable spokesperson in a striped polo shirt with a Green Bay Packers sticker on his hard hat, served as our guide as we ventured into the mineral deep. Seconds later, the walls around us turned from industrial concrete to exposed crystal as we pierced a three-thousand-foot-thick underground salt deposit left behind by an ancient sea. The acoustics of our journey—dropping half a mile straight down into the planet—changed dramatically. We could hear one another speak again.

The Waste Isolation Pilot Plant, or WIPP, is the world's first—and, as this book goes to press, only—active facility for the permanent, deep geologic disposal of nuclear waste. WIPP is located 2,150 feet below ground in the New Mexican desert, roughly thirty miles east of Carlsbad, near the border with Texas, inside a piece of land the Department of Energy refers to as "the Withdrawal." The name is a reference to a 1992 Land Withdrawal Act that transferred the sixteen-square-mile area from New Mexico state control to the long-term supervision of the DOE and the Environmental

The elevator at the U.S. Department of Energy's Waste Isolation Pilot Plant (WIPP) outside Carlsbad, New Mexico, drops workers and visitors alike into a large hall carved from a 250-million-year-old salt deposit. The elevator ride alone takes several minutes, traveling 2,150 feet into the Earth. (*Photograph by Nicola Twilley*)

Protection Agency. The outlines of the Withdrawal—a perfectly square piece of land, in the very center of which sits WIPP—become more clearly delineated in satellite photos every year as oil pumpjacks move as close to its edge as they legally can without violating the facility's safety protocols.

The entire region around WIPP is something of a subterranean wonderland. Carlsbad Caverns National Park is nearby, whose attractions include the stalactite-festooned chambers of Carlsbad Cavern itself, as well as the mysterious Lechuguilla, once a candidate for the world's deepest cave and home to a collection of unusual microbes that makes it a site of intense interest for researchers exploring the extremes of terrestrial life. Whether because of show caves, sinkholes, potash mines, fracking wells, or the burial of radioactive waste, this is a region that exists in an ongoing exchange with its subsurface.

Although WIPP is not a quarantine facility—as its name states, it is a place of *isolation*—it is perhaps the ultimate Earth-based destination for any quarantine tourist keen to see extreme engineering controls in action. It is a burial site, designed to keep potentially hazardous materials isolated for at least ten thousand years—a radioactive tomb, containing "a kind of waste that resists its own containment," as Peter C. van Wyck, a scholar in the field of disaster studies, has memorably phrased it.

Nevertheless, the materials buried at WIPP are not warheads from old atomic bombs or the cores of dismantled reactors. The facility instead stores what John Howard, in his own time, described as "susceptible matter"—or, as St. John explained to us, "clothing, tools, rags, residues, debris, soil, solidified sludge, and gravel," all of it contaminated with small amounts of plutonium and other man-made radioactive elements.

These radioactive elements, heavier than uranium, are called *transuranic*, or, literally, "beyond uranium." They are also all artificial, created not by natural processes but by human industrial activity—specifically, in the case of the material destined for WIPP, during the development, maintenance, and testing of the nation's nuclear arsenal. "Transuranic waste is about a thousand times less radioactive than spent nuclear fuel and generates much less heat," William Alley and Rosemarie Alley write in their book *Too Hot to Touch*. "Most transuranic waste can be handled when properly stored in containers. The main problem is its *longevity*."

Time is a trickier problem than space in this particular configuration of containment—indeed, potentially an insurmountable one. Much of the material buried at WIPP will remain dangerous to living organisms not for the traditional forty days of quarantine but for at least ten thousand years—some of it for many millions of years—posing an extraordinary challenge for anyone hoping to design an effective form of isolation. The timescales required for the safe disposal of transuranic waste are, in fact, so daunting that the architects of a deep geologic repository in Finland known as Onkalo—expected to finish construction and begin receiving nuclear waste in the 2020s—have had to account for the ebb and flow of future ice ages. All of northern Europe might be buried beneath mile-high glaciers not just once but several times before Onkalo's radioactive payload can be considered safe.

At an equivalent facility in Nevada, known as Yucca Mountain—intended for the interment of spent nuclear fuel, it was defunded by the Obama administration in April 2011, though its future regulatory reapproval seems possible—the federally mandated period for the containment of nuclear waste was an astonishing *one million years*. One million years ago, *Homo sapiens* did not yet exist on Earth, and it is all but impossible to predict whether humans—let alone U.S. regulatory agencies—will still exist one million years from now, when the contents of Yucca Mountain will, at long last, be considered safe for biological exposure. Constructing a spatial system to isolate anything for spans of time this immense is so quixotic in its ambition that rational engineering reports and analyses become indistinguishable from science fiction.

At one point, while attending a nuclear-waste conference in Phoenix, Arizona, we watched as one participant became so flustered by the complexity of the problem of long-term containment—its billion-dollar expenditures, its requirements for thinking hundreds of generations into the future, its necessity for engineering entirely new classes of materials capable of resisting earthquakes, floods, and radioactivity—that he finally stood up and asked why we couldn't just take all this valuable time and money and use it to cure cancer. Problem solved, the man suggested: we could just vaccinate ourselves and happily spend a lifetime exposed to spent fuel rods. (Countering the physiological effects of radiation poisoning, alas, is not simply a matter of curing cancer.)

As it is, nuclear waste around the world now sits in a strange kind of purgatory. Contaminated soil is held in "interim" storage near the ruined reactor in Fukushima, Japan, relying on tens of thousands of waterproof bags; in Hanford, Washington, highly radioactive liquid waste awaits permanent disposal inside enormous, leak-prone underground tanks, some of which are nearly sixty years old. Most of the material currently destined for WIPP lurks in aboveground casks, vulnerable to corrosion, natural disaster, and terrorist attack. No government—and no credible nuclear engineer—believes that this is safe or sustainable. Facilities such as WIPP are a potential answer: precision graveyards constructed to deal, once and for all, with the problem of anthropogenic radioactivity.

Walking inside the corridors of WIPP, known as "drifts."
(*Photograph by Nicola Twilley*)

Our four-minute-long plunge into the Earth complete, we stepped off the elevator into a hot, dimly lit corridor roaring with ventilation ducts, the salt-crystal walls on either side strung with electrical wires. Our feet crunched across the carved-salt floors with a sound like boots on fresh snow, and our lips tasted salty. The air reeked of diesel exhaust coming from huge pieces of industrial equipment that lumbered away in a storm of headlights and reverse-gear warning beeps ahead. Roughly five hundred feet away from the elevator shaft, another corridor branched off, its salt walls matte with dust, leading to another, and another, in a labyrinth of identical corridors, officially known as "drifts." The drifts connect onward to a total of eight "panels"—not all of which have yet been excavated, although six entire panels are already full—and each panel contains seven "rooms." WIPP, to put it mildly, is huge. The so-called North Experimental Area—a quarter again as large as the rest of WIPP—has been set aside for scientific research, including underground plant-growth studies. Current estimates are

We drove deep into the facility with Bobby St. John. The walls and ceilings are slowly collapsing under the weight of the Earth above. "The rock never stops," WIPP's chief scientist, Roger Nelson, told us later that day. (*Photograph by Nicola Twilley*)

that, by 2033, the facility will hold enough nuclear waste to fill more than one hundred basketball courts, at which point WIPP will be declared full. From that point on, the mine will be permanently sealed, isolated from the surface of the Earth, its architects hope, forever.

The walls and ceilings around us, carved into the salt as long ago as the 1980s, were sagging, often dramatically, their once-straight lines now bulging, held back from collapse only by steel nets, rock bolts, and geotechnical pins. This closure rate, as it's known, is fully five inches per year. "The rock never stops," WIPP's chief scientist, Roger Nelson, told us later, explaining that the ongoing excavation of new rooms and panels has to be timed according to the arrival of future waste. "The depth underground, plus that closure rate, plus the rate at which we expect to be able to ship a zillion drums a year to WIPP—it all balances out to the point where we can keep the rooms open, and, by the time they're ready to become a safety issue, they're full. It's a balancing act between the geology and the waste-receipt rate." WIPP is, in this sense, a contradiction in terms: a just-in-time architecture for eternal isolation.

Counterintuitively, WIPP's geological instability is one of the reasons why this site was chosen in the first place. Under enough pressure—such as being weighed down beneath more than two thousand feet of Earth, every second of every day—salt behaves a bit like marshmallow, or, as Nelson described it, "molasses in January." It bulges, oozes, and creeps. Long after WIPP is closed—long after all those geotechnical pins and steel nets have failed—the salt will flow together like a closing wound, collapsing around the radioactive waste containers and crushing them into a permanent crystalline tomb. Approaching the ominously bulging salt, we unclipped safety lights from our hard hats and held them up against the wall: the light penetrated more than a foot into the crystalline mass, creating a yellowish glow that gave the entire place an ethereal quality, its walls translucent and lit from within.

Initially confused by a howling breeze that ruffled our clothing as we walked down the corridor, we recalled the safety briefing we had been given before our descent. The facility's underground air system is designed, we were told, to move air in only one direction: away from the elevator shaft and deeper into the drifts. There, the air can be filtered, vented, and cycled back out again. This means that, in the event of an accident or radiation leak, we could find our way out of WIPP by heading into the wind, following the subterranean breeze back to safety.

Just in case, we had also been equipped with a "self-rescuer"—a portable respirator designed to supply enough oxygen for us to escape in the event of a fire—and dosimeters to monitor ourselves for exposure to radioactivity. Readings from our dosimeters would be mailed to us in a few weeks, but we were told not to worry; it was a routine precaution. With that, we walked over to an electric golf cart charging in a side chamber, hopped on beside St. John, and began the tour, the wind to our backs as if sailing into the deep.

⟷

The Waste Isolation Pilot Plant, we learned from our primary guide to the facility, the DOE geophysicist Abraham Van Luik, is a demonstration facility. As its name implies, it is a *pilot* plant—an experiment. WIPP's construction was authorized by an act of Congress in 1979, after the successful submission of a one-hundred-thousand-page environmental plan, but it took another two decades of heavy construction and mechanical preparation before the first shipment of transuranic waste could be delivered, in March 1999.

At the time of our tour, we had been in touch with Van Luik for several years. (Sadly, Van Luik died in 2016, at the age of seventy-one, mere months after our last meeting with him.) When we first spoke with him, in 2009, he was not working at WIPP but in southwest Nevada, helping to conceptualize and engineer the even-more-extreme nuclear-waste disposal site at Yucca Mountain. Yucca Mountain, a massive landform created by an extinct supervolcano, lies ninety miles northwest of Las Vegas on federal land adjacent to the secretive Nevada Test Site, where aboveground nuclear weapons tests had been conducted throughout the Cold War. In 2002, it was chosen by Congress as the national repository for radioactive waste from nuclear power plants all over the United States—including, but not limited to, spent fuel rods now languishing inside huge pools of water used to cool the hot material. Should this water ever evaporate completely, the fuel rods themselves risk catching fire, releasing lethal radioactive plumes; without a long-term solution, this could happen at any moment for thousands of years to come. Delivering such exceptionally dangerous waste to the facility would require a long and complex exercise in quarantine logistics: a journey, by rail and road, through dozens of intermediary states whose politicians were not at all enthusiastic about such a prospect.

In a report released in August 2001 called "Worst Case Credible Nuclear Transportation Accidents," consultants hired by the DOE outlined a handful of "potential severe accident scenarios" that might occur while shipping high-level waste, such as spent nuclear fuel rods, to Yucca Mountain. In the authors' analysis, the worst of the "most credible severe accidents" would result from a high-speed collision near downtown Las Vegas; the authors outline possible truck- and rail-based incidents.

The resulting radioactive plume from a broken waste

cask would almost certainly contaminate huge swaths of the city. At least 138,000 people would be immediately affected and the city's buildings, including major international convention and hotel facilities, could become reservoirs of radioactivity. "If ventilation systems were not shut off," the authors write, "radioactive particulates would settle within hotels and other buildings, contaminating rugs, furniture, beds." Worse, if the ventilation systems *were* shut off—but too late (the report's mathematical model gives nearby hotels only 91.3 seconds before they are enveloped by the resulting radioactive plume)—those same particulates would be trapped there, irradiating building interiors beyond salvage. Radionuclides that settled onto the city's soil, plants, and streets would also emit a dose of gamma radiation, technically known as *groundshine*.

Radioactive particles would likely be fanned outward from Las Vegas by cars and trucks; the study recommends shutting down the city's international airport, as well, in order "to prevent the migration of contaminated persons" who might inadvertently spread radioactive particles around the country and world. "Given the high number of people exposed, local responders will not be able to identify, let alone effectively quarantine, contaminated people," the report warns.

In the end, saving the city might not be possible; the authors calculate that one decontamination technique, known as *sponge blasting*, "would take nearly 25,000 machine-years" to achieve safety. Much of Las Vegas—down to its very streets and sidewalks—might have to be razed. The authors' proposed alternative to a complete dismantling of the city is what they call "permanent quarantine," by which they effectively mean abandoning Las Vegas to the sands.

The material transported to WIPP does not pose these same kinds of existential threats, Van Luik assured us. Nevertheless, the logistical challenge of containing even low-level wastes—gloves, gowns, and laboratory equipment potentially exposed to radiation—should not be underestimated. Bobby St. John told us that much of his time is spent conducting outreach along potential new trucking routes, meeting with elected officials and first responders. To qualify to drive for WIPP, candidates must undergo a background check and have no traffic violations in the past seven years: drivers work in teams so that, unless they encounter a severe weather alert en route, the waste never stops moving. If necessary, the DOE will pay to upgrade the roads themselves: St. John told us that federal funding had recently transformed US 180 from a two-lane road into a four-lane divided highway west of Carlsbad. If you have driven through New Mexico, in other words, you have likely, without knowing it, used a highway bypass that exists because of WIPP; the much-mythologized open road of the American West is also part of an infrastructure of geologic isolation, hiding in plain sight.

Not long after our first conversation with Van Luik, in 2009, the entire Yucca Mountain facility was closed by the DOE, and Van Luik transferred to New Mexico, joining WIPP. When we met him in person there, we found him both avuncular and kind, two qualities that might seem unexpected for a federal geophysicist working in nuclear-waste disposal, but Van Luik was a man of several surprising facets. On his personal blog, for example, he often wrote about Catholic mysticism, including the tension between intellect and intuition. Van Luik framed these as incompatible aspects of the human condition that needed to be held apart from each other, one buried just beneath the surface as if in a state of quarantine. "I will live with this interior division in myself because it allows me to live in two worlds at the same time," Van Luik wrote; this separation gave him space to work as a geophysicist, his rational scientific rigor uncontaminated by a belief in the afterlife or a personal relationship with the divine.

As it happened, he was also a passionate fan of Dante's *Divine Comedy*, enamored with the poem's descriptions of "transcendental love" between Dante and his muse, Beatrice. In Dante's story, it is this love that helps to pull the poem's narrator out of the subterranean horrors of the Inferno and onto the rocky slopes of Purgatory, where souls wait, until proven worthy, to be set free in Paradise.

Underworld metaphors abound in the global nuclear-waste industry, just as human mythology is filled with stories of monstrous things locked in the deep. The HADES underground research facility in Belgium, for example, takes its name from the Greek word for Hell; HADES, in this case, stands for High Activity Disposal Experimental Site. HADES, in turn, is run by a consortium known as EURIDICE, named after Eurydice, whose husband, Orpheus, unsuccessfully attempted to rescue her from Hell. (Incredibly, EURIDICE is an acronym for European Underground Research Infrastructure for Disposal of Nuclear Waste in a Clay Environment.)

Last but not least, CASTOR, or Cask for Storage and Transport of Radioactive Material, is the name used by the European nuclear industry for its temporary or interim waste containers; the ones used for permanent disposal are named Pollux—which, disappointingly, is not an acronym. Castor and Pollux, of course, were twin brothers in Greek mythology; Castor was mortal (or, as it were, temporary) and Pollux immortal (or permanent). In the twins' story, facing the prospect of spending eternity apart once it is Castor's time to die, the brothers instead make a deal with Zeus: they will split immortality between themselves, spending the rest of all time alternating between Mount Olympus and Hades, or Heaven and Hell, Paradise and the Inferno.

In his role as a DOE spokesperson, Van Luik avoided the mystical, sticking instead to technical discussions of the advantages and disadvantages of different geological barriers and describing each step in the process by which robots are now being used to help fill the repository. As we would see, remote-handling systems are used for the most-radioactive waste, removing humans from the equation almost entirely in an elaborate choreography of separation, all the way up to the final moments before the waste is lowered into the salt tombs below.

Like John Howard, who once dreamed of how he might design the perfect lazaretto, Van Luik has imagined his own ideal facility for isolating nuclear waste. "My ideal repository location has changed over time," he explained. "When I worked on crystalline rock, like granites, I thought crystalline rock was the cat's meow." Van Luik briefly flirted with salt, as found at WIPP, but, he admitted, "Now that I have worked with the European countries and Japan for the past twenty-five years, learning of their studies of various repository locations, I'm beginning to think that clay stone is probably the ideal medium."

Van Luik emphasized that Yucca Mountain and WIPP are just two particular examples of how humans might dispose of nuclear waste, and that other nations continue to experiment with other strategies—and other geologies. China, for example, has recently taken the first steps toward constructing its own permanent disposal site, excavating an experimental test facility—similar to WIPP—in deep desert rocks near Beishan, in Gansu Province, in the middle of the Gobi Desert. (Beishan lies near the Silk Road, the old trade route traveled by Marco Polo and other merchants—some of whom carried fleas infected with bubonic plague to Europe, thus prompting the very first architectures of quarantine.)

Meanwhile, some engineers argue in favor of not building repositories at all, instead sinking sealed capsules of nuclear waste directly into the abyssal muds of the open ocean floor; others are looking at locking nuclear waste inside huge, impenetrable glass bricks the size of shipping containers using a process called vitrification; yet others are hoping to use radiation-consuming microbes to help convert buried waste into less dangerous, more stable forms. According to the industry's true optimists, nuclear waste is not waste at all but an untapped source of energy that might someday fuel ultra-long-life batteries for prosthetic medical devices and spaceships. For now, the afterlife of our nuclear activities remains both an unsolved challenge and one that will haunt all of our descendants for thousands of generations to come—a responsibility that weighed heavily on Van Luik.

⟷

Our own journey into the world of deep-geologic disposal was inspired by a desire to see how ambitious infrastructures of isolation, operative over geological timescales, might reframe our understanding of quarantine. Although we had come to WIPP to see the outermost limits of humanity's containment capabilities, we quickly realized that—as with quarantine—the purely *technical* challenge of isolation was not the most formidable hurdle.

The key to successfully keeping something dangerous isolated from the rest of the world for any significant period of time is communicating the necessity of that separation to other humans. Warning people that they are at risk and making sure that they believe you is fundamental to all attempts at isolation and quarantine. This became all too evident during the COVID-19 pandemic, when the United States suffered a disproportionate death toll in large part because so many of its citizens simply did not believe medical experts who said that the coronavirus was a threat; some Americans did not even believe it was a real virus. Recall, as well, that, throughout the history of quarantine, warning signs that a particular house had been infected were sometimes interpreted by the greedy, or simply opportunistic, as a burglary targeting system: those were houses not to avoid but to break into.

When the goal is to communicate similarly existential risk to people in the unimaginably distant future, the problem becomes all but insuperable. "There is no international standard on how to warn future generations," Van Luik told us. Nonetheless, places such as WIPP, Finland's Onkalo, and—if it is ever reopened—Yucca Mountain have an ethical responsibility to convey how dangerous their contents are to future humans living tens or even hundreds of thousands of years from now. To put this into historical context, the inner shrine of the tomb of King Tutankhamen in Egypt was left intact little more than three thousand years before it was opened and its contents removed. WIPP, by contrast, will need to remain undisturbed for at least three times as long in order to meet federal regulatory goals. Indeed, everything that historians currently recognize as human civilization has occurred in the last ten thousand to twelve thousand years; WIPP requires that same amount of time all over again for its contents to achieve even minimal physiological safety.

Van Luik described the resulting dilemma to us in terms of four categories. "A message has to survive, be found, and be understood," he said. "It also has to be believed—that's the hardest part." Warning signs meant to deter future archaeologists or industrial salvage crews from breaking into WIPP must, of course, be *durable*: the signs must be physically capable of surviving for millennia without simply eroding or rusting away. This has inspired experiments with high-grade industrial materials, including corrosion-resistant metal alloys. In the end, massive pieces of what the Department of Energy calls "pristine granite," each block weighing as much as sixty-five tons, have been selected for WIPP's final markers. In an appendix attached to the final WIPP application for regulatory approval, the DOE actually boasts that "the weathering and erosion characteristics of this granite are superior to the rock types (silicified sandstone and dolomite) employed in the construction of Stonehenge."

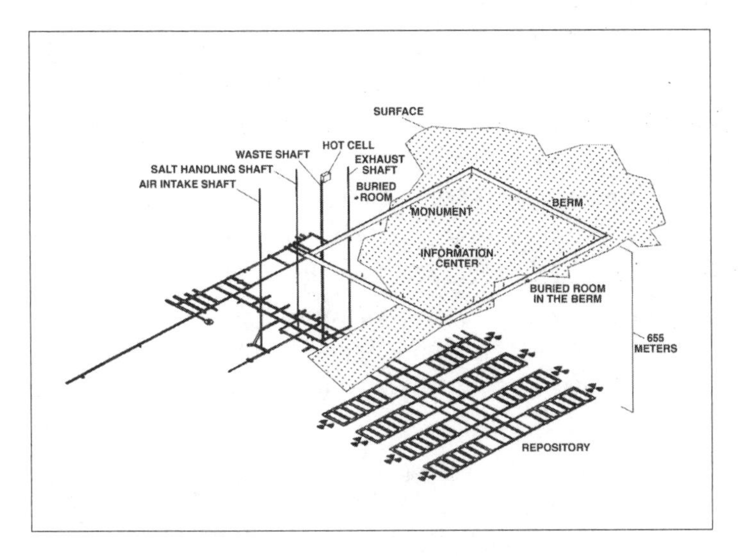

After WIPP receives its final load of nuclear waste—currently estimated to be the year 2033—it will be surrounded by a huge earthen berm and a system of large granite markers, seen here. (*Courtesy of the U.S. Department of Energy*)

These durable warning signs must also be *locatable*, Van Luik continued. We cannot risk them being lost to history somewhere, like the Dead Sea Scrolls, awaiting accidental rediscovery by errant nomads. Part of WIPP's assurance for this is the future construction of an earthen berm: at 33 feet high and 98 feet wide, requiring 975,000 cubic yards of material, this huge artificial landform will enclose—and thus help to mark—the entire WIPP site. As an added benefit, the berm will also act as a wind deflector, helping to protect the granite markers inside from erosion.

Raising the stakes yet higher, these durable, easy-to-locate granite warning signs must also remain *legible* and *comprehensible* for tens of thousands of years. They cannot share the same fate as Linear A, for example, an ancient Minoan writing system from roughly 1800 BCE that remains undeciphered to this day.

In an April 1984 paper titled "Communication Measures to Bridge Ten Millennia," the Indiana University semiotician Thomas A. Sebeok proposed a number of different solutions to the challenge of long-term comprehensibility. The ultimate goal, Sebeok wrote, "is to devise a method of warning future generations not to mine or drill at that site unless they are aware of the consequences of their actions." Although it was only one of many such papers produced by a DOE-commissioned group known as the Human Interference Task Force, Sebeok's paper has become something of a cult classic in speculative-design circles.

Among the many ideas put forward by the DOE's research group were several that, in retrospect, seem so ill-considered as to appear frivolous. One such proposal suggested building a "landscape of thorns"—fifty-foot-tall concrete spikes—on the land above WIPP, under the assumption that large, geometrically aggressive shapes would be so menacing that anyone who saw them would simply flee in horror. Another called for the genetic modification of domestic house cats in order to produce a new species, the "ray cat," whose skin would change color in the presence of radiation, possibly even emitting fluorescent light. Many thousands of years from now, the thinking behind this suggestion goes, humans passing through lands formerly known as New Mexico would notice that their cats were changing color, and they would conclude from this that radiation must be leaking from a long-forgotten underground repository. A place of safety would be anywhere their cats no longer glow.

An equally baffling proposal suggested reproducing, throughout the WIPP region, an illustration inspired by the Norwegian Modernist painter Edvard Munch's *The Scream*. Despite the painting's popularity as a refrigerator magnet and its eventual use as a comedic reference in the poster for Macaulay Culkin's 1990 film *Home Alone*, *The Scream* was considered to represent such elemental terror by Department of Energy semioticians that even humans living several thousand years from now will simply leave the area rather than gaze upon the fearsome contours of Munch's work.

Yet, somehow, it gets worse. We were taken aback to hear, while attending the nuclear-waste conference in Phoenix, a then employee of the DOE propose his own idea for warning future generations: WIPP should create an exceptionally detailed legacy website, accessible by smartphone anywhere in the world, complete with descriptions, images, and plans of what lies below. It would be called the World Information Library. The possibility that smartphones, let alone Wi-Fi, let alone the internet, might not exist five thousand years from now did not seem to cross the man's mind.

Sebeok himself proposed the establishment of an "atomic priesthood," "a commission of knowledgeable physicists, experts in radiation sickness, anthropologists, linguists, psychologists, semioticians, and whatever additional expertise may be called for now and in the future." This atomic priesthood would keep the memory of WIPP alive through the use of narrative rituals and what Sebeok calls "folkloristic devices." New myths and legends extolling radioactive horror should be devised, Sebeok suggested, every three generations for at least ten thousand years. "Essentially," he writes, the atomic priesthood will cultivate an intergenerational air of "superstition," one that will cause people "to shun a certain area permanently." Of course, the history of Christianity alone suggests that such a project is untenable: the schisms, heresies, inquisitions, reformations, and even cults that have entered the interpretive fray over the past two thousand years do not bode well for a project that needs at least ten thousand years before the dark materials it warns about are proven safe.

The most difficult challenge of all, Van Luik told us, is that, no matter what form these warnings ultimately take, they must remain *credible*. Future generations must take the U.S. Department of Energy at its word that nuclear waste is dangerous and that they really—truly, seriously—should not exhume it. An oft-used example here are the inscriptions found—and immediately ignored by European archaeologists—outside the tombs of Egyptian pharaohs, instructing would-be trespassers not to violate these sacred burial sites or else suffer the consequences. These warnings were dismissed at the time as ancient superstition, but what, then, will future humans make of nearly identical warnings carved by twenty-first-century geophysicists claiming that something of immense danger is hidden below ground with no safe way to dig it up? Sebeok's explicit reliance on folklore would only seem to guarantee that such warnings will be dismissed as a particularly macabre kind of pagan belief.

In the DOE's current philosophy, informing future generations that a repository's contents are both dangerous and useless makes more sense than trying to scare those people away. As a line taken from one of the DOE's own suggested warning messages, potentially to be carved onto these future granite monuments, explains, "This is not a place of honor . . . no highly esteemed deed is commemorated here . . . nothing valued is here." As Van Luik himself noted, any repository project necessarily incorporates a massive amount of scientific and industrial materials that are considered very valuable today, albeit too dangerous to use. In fact, while WIPP's name emphasizes its role as an experiment in isolation, it also draws attention to the concept of *waste.* One of the technical problems anticipated by the facility's designers was the possibility that the materials entombed at WIPP will someday be considered anything *but* waste, seen instead as valuable commodities to salvage—and thus, for future humans, worth the risks involved in attempting their retrieval.

We were reminded of health regulations governing the burial of potentially infectious corpses during the Black Death. Venetian officials required that plague victims be placed in graves, rather than in aboveground tombs, as soil and earth were seen as an extra layer of protection from future outbreaks. Like WIPP, these burial sites were to "remain undisturbed for a period of time deemed sufficient by the Health Office," the historian Jane Stevens Crawshaw writes, in order for their contents to be considered safe. "Otherwise, the doctors noted, opening up the graves would be like the opening of the gold casket in the Temple of Apollo at the time of Avidius Cassius when the soldiers expected to find treasure but instead met their death because a pestilential vapor was released." Even today, Mike Jacobs of the Royal Free Hospital told us, the bodies of Ebola victims are sealed inside specially welded zinc caskets before they leave the isolation unit, lest mourners be tempted to open them—and release their contagion.

Deliberate, or "advertent," retrieval scenarios are explicitly described in WIPP's own technical application documents. Advertent entry is considered all but impossible to prevent—and is, in fact, deemed so likely to occur that, in their book on "principles and standards" for the geologic disposal of radioactive waste, the engineers Neil Chapman and Charles McCombie suggest that facilities such as WIPP, Yucca Mountain, and Onkalo will inevitably be breached. "It is interesting to consider that no unusual work of ancient man has ever been left wholly undisturbed once it has been discovered," Chapman and McCombie write, concluding that "the probability that a deep repository will ever be allowed to demonstrate its long-term containment capacity may be low."

There would be considerable irony in our having gone to such lengths to dispose of these materials only to discover that, centuries or millennia from now, what we so carefully buried has accumulated enough value for our descendants to dig it all back up again. The afterlife of materials contaminated by the Chernobyl nuclear power plant explosion in 1986 offers an instructive example. The nearby city of Pripyat, Ukraine, was abandoned, and the entire area, deemed too radioactive for human inhabitation, thereafter became known as "the Zone." Within a few years, however, the Zone's now-derelict buildings—old schools, factories, offices, and even private homes—had been stripped of nearly six million tons of valuable but radioactive metal, which was believed to have been sold off to scrap retailers throughout Europe. (In January 2012, *Bloomberg* reported that American retail chain Bed Bath & Beyond had unwittingly been selling radioactive metal tissue boxes, raising "alarms among nuclear security officials and company executives over the growing global threat of contaminated scrap metal.") Where an opposing economic incentive exists, quarantine and isolation are all but impossible to enforce.

Given Van Luik's other interests, another famous warning inscription came to mind. In "The Inferno," the first book of *The Divine Comedy*, Dante describes a small sign carved above a rocky gate that leads deep into Hell. *Lasciate ogni speranza, voi ch'entrate*, the inscription warns—or *abandon all hope, all ye who enter here*. It is not out of the question that future excavation teams or industrial drilling crews will simply laugh off such warnings, crank up their rigs, and continue digging farther into the Earth.

⟷

Rolling along in our open-top golf cart, its headlights barely penetrating the subterranean gloom, we came to a series of massive, pneumatically powered doors that sealed one part of the mine from another. At each door, Bobby St. John had to stop, reach upward, and grab a dangling tube; after he pulled the tube down like an airhorn, the huge, eerie doors in front of us, sized for industrial earthmoving equipment, hissed open with a sound not unlike the exhalations of Darth Vader.

After passing through several such doors and traveling down corridors so long we could not see the end of them, we finally reached the "waste face," as it is called: nuclear waste awaiting its slow, salty embrace. St. John parked the cart over to one side and we hopped off to get a closer look, nervously double-checking each other's security IDs to ensure that our dosimeters were still attached.

Not more than thirty feet ahead, behind a waist-high yellow chain strung from one side of the corridor to the other, was a wall of industrial drums and specially designed casks. They were simply stacked there, like forgotten items in a storage unit. Clusters of metal barrels, we saw, had been wrapped together using cling film. A small sign dangled from the yellow chain warning: CAUTION—RADIATION AREA. It was a strangely anticlimactic sight, given the mythological buildup; seen out of context, these informally

Industrial drums and specially designed casks sit behind a yellow chain with a radiation-warning sign. This is the "waste face." (*Photograph by Nicola Twilley*)

grouped barrels and plastic-wrapped packages could pass for cargo being loaded onto an airplane. (Other, more high-risk wastes, we'd soon see, are instead inserted into cylindrical hollows that have been cored horizontally into the walls; the waste package is then slipped inside, in a process that resembles loading a bullet into a revolver, and the cavity sealed with a small lid.)

Although the sight of stacked containers behind a chain was far from the terrible glowing mass we had imagined, this mismatch between expectation and reality raised the question of what risk and danger are supposed to look like in the first place. Radioactivity, like a virus, is invisible to the naked eye. We thought of the seventeenth-century quarantine guard in Split who saw a beautiful scarf, took it home after a long day of work at the lazaretto, and gave it to his wife, unwittingly unleashing a plague on the city.

It was a long ride back to the elevator shaft, driving down vast, slowly collapsing corridors filled with the relentless din of mechanical ventilation. Heading back into the wind—back toward safety—we sat, awed and slightly stunned, attempting to reckon with the inhuman scope of our surroundings.

Sites of containment, isolation, and quarantine—whether they are animal-disease labs, plant-research facilities, underground nuclear-waste repositories, or emergency medical wards—are direct outgrowths of how we model and understand uncertainty. Places such as WIPP, Yucca Mountain, the National Bio and Agro-Defense Facility, and even John Howard's ideal lazaretto are what results when abstract arguments over risk are given architectural form. They are where proximity to danger becomes spatial and philosophical at the same time. The cavernous, howling space we toured that day was, in effect, the inevitable by-product of what it means to seek ten thousand years of certainty. Panel by panel, room by room, it was as if WIPP had grown in the darkness, expanding in time with our fears of potential exposure.

The lesson of WIPP, in its current incarnation, seems to be that if we create and fund a federal project meant to ensure that a pair of laboratory gloves, potentially contaminated with trace amounts of radiation, has no chance of harming living creatures at any point in the next ten thousand years, then it makes sense to design and build a sprawling system of transportation, packing, and burial that becomes excruciating in its ritualistic thoroughness.

While we were touring the Royal Free Hospital in London—following a carefully marked, one-way route designed to ensure that no infected materials could make their way back into a space considered safe—Mike Jacobs told us that, while many might see the High Level Isolation Unit's elaborate precautions as overkill, they reflect a different calculus. "The principle here," he said, "is not risk assessment but consequence assessment. That's a very, very important distinction. Undoubtedly, the *risk* to the U.K. of diseases like Ebola is very low—but the *consequence*, if you get it wrong, is very substantial. So that is the basis on which this unit exists."

Any true *zero*-risk architecture requires thinking through endless chains of contingency, from industrial accidents and natural disasters to deliberate acts of sabotage: a million linked uncertainties about what might happen, how, when, whether or not that event can be prevented, and what to do when it occurs. This includes so-called low-probability, high-impact events—like a nuclear fire on a Las Vegas freeway. Such a thing will, almost certainly, never happen—but, if it does, we could lose Las Vegas.

Whether it is a material, such as nuclear waste, a virus, such as Ebola, or even a fungus, such as wheat rust, if you are told that something has to be isolated with zero chance of exposure, potentially for thousands of years, then you are posing a challenge that, by necessity, will result in systems on top of systems, stopgaps behind stopgaps. You will need nested defenses and always-ready alternative plans that may well defy the human imagination, but will certainly defy the budgetary limits of anything other than a well-funded nation-state.

It was thus comically disconcerting to learn that, despite all of WIPP's precautions, the whole thing was nearly undone by kitty litter. On Valentine's Day 2014, a "radiological release event" occurred underground at WIPP when a drum filled with transuranic waste shipped from the Los Alamos National Laboratory partially exploded. The drum—number 68660—had been packed using the wrong kind of kitty litter. Instead of a clay-based litter—whose composition has the beneficial side effect of helping to block radiation—workers used an organic litter, one comprised of a "wheat-based absorbent." This seemingly minor decision had terrible consequences, leading to a chemical reaction that, according to an April 2015 incident report issued by the DOE, triggered "pressurization of the drum, failure of the drum locking ring, and displacement of the drum lid." In other words, an explosion—one that did an estimated $500 million worth of damage.

Trace levels of the radioactive elements americium (whose half-life is 432 years) and plutonium (with a half-life of 24,100 years) were later detected beyond the outer bounds of the Withdrawal, having been accidentally vented from below by the facility's powerful air-handling equipment. Like all examples of isolation and quarantine, this breach in the armor seems so trivial and mundane as to be hard to fathom how it could have happened in the first place. Nevertheless, a retired nuclear chemist who helped to determine what really occurred inside this particular canister later admitted that it might well happen again: nearly seven hundred drums, he pointed out, all buried at WIPP, were packed using this same organic kitty litter. Why only one of them has blown up so far remains unclear.

⟷

A few weeks after our visit to WIPP, two envelopes from the Department of Energy arrived at our home: our dosimeter results had come back. Feeling both apprehensive and excited, we ripped them open in our kitchen. We had been exposed to no radioactivity.

8

All the Planets, All the Time

"This—what you're doing today—never happens," NASA's David Seidel told us. "This is a rare chance," agreed the director of the Jet Propulsion Laboratory (JPL), Michael Watkins, welcoming us to the lab's Spacecraft Assembly Facility, located just a few miles north of the Rose Bowl, in the hills outside Pasadena, California.

The exceedingly unusual adventure awaiting us was a trip into the clean room where *Perseverance*, NASA's latest Mars rover, having been assembled under conditions of exacting sterility, sat awaiting shipment to Cape Canaveral. Our visit had been prefaced by a long email laying out extremely detailed rules: we were instructed not to wear any perfume, cologne, makeup, or dangly earrings; flannel, woolen, or frayed clothing was not allowed; even our fingernails had to be smooth, rather than jagged.

After a quick welcome, our phones and notebooks were confiscated, and a high-tech doormat vacuum-brushed the soles of our shoes. In the gowning room, we were issued with face wipes, a sterile full-body "bunny suit," Tyvek bootees, hood, gloves, and face mask, then offered a mirror in which to admire the final look. Finally, we were sent through the air shower—an elevator-size chamber studded with nozzles that blasted us with pressurized air from all sides, in order to dust off any final stray particles—before stepping out into a white-floored, white-walled room filled with white-suited engineers.

The rover itself—a white, Rube Goldberg go-kart the size of an SUV—was cordoned off behind red stanchions. The obsessive attention to cleanliness required in order to enter the rover's presence was, in part, to protect the machine's sensitive optical equipment and electronics: volatile chemicals, loose fibers, and even flakes of human skin could damage its delicate circuitry or settle on one of its twenty-three camera lenses. But the primary purpose was planetary quarantine: preventing the importation of Earth life to Mars. "I don't know that we can say it's the most sterile object that humans have ever created," said one engineer. "But it's extremely clean."

It is a Faustian condition of space exploration that we cannot search for life on alien planets without bringing along very small amounts of very small Earth life. This process is known as forward contamination, and minimizing, if not preventing, it is the ultimate responsibility of NASA's Planetary Protection Officer—"the second-best job title at NASA," according to its previous holder, Catherine Conley. (Conley, who goes by Cassie, served in the role for more than a decade, from 2006 to 2018, when she was succeeded by Lisa Pratt.) The best job title, according to Conley, was Director of the Universe, but that position was sadly eliminated in an institutional reorganization.

The practice of planetary quarantine dates back to the 1950s, when it became clear that rocket technology was shortly going to put outer space within human reach for the first time. In an ideal universe, the robotic spacecraft that we send to explore the cosmos would be sterile. (Humans are, by definition, contaminants.) In reality, for both technical and economic reasons, they are not. But the consequences of transferring biological material between celestial bodies are a fractal example of unknown unknowns: we don't know what forms of Earth life might survive a space journey, which of them might then flourish in whatever extraterrestrial conditions await them, and whether life even exists elsewhere in the solar system, let alone how it might be harmed by Earth life—or vice versa.

Faced with such extreme uncertainty, but unwilling to stay at home, spacefarers, like so many before them, have turned to quarantine as the buffer that will allow them to explore space responsibly, without endangering Earth or inadvertently polluting the cosmos. In this context, quarantine is an elaborate set of protocols designed to allow the reduction of that uncertainty, while minimizing the risk that exploration might cause accidental, irrevocable damage. If animal and vegetable biosecurity lays bare the unsparing calculus of quarantine, in which life's value is always subject to economic imperatives; and mineral quarantine exemplifies the surreal megastructures that result from the quixotic pursuit of total containment; then planetary quarantine is the impossible art of modeling risk when your data is nonexistent but the stakes are existential.

⟷

The woman in charge of protecting "all the planets, all the time," as international planetary protection policy puts it, works out of a small office inside NASA's headquarters, a squat, undistinguished building in Washington, D.C.

Just a couple of blocks to its north, in between the U.S. Botanic Garden and the Air and Space Museum, lies a fairly recent addition to the Mall: the National Museum of the American Indian. It was established in response to the controversial revelation that the National Museum of Natural

History held the skeletons of nearly twenty thousand Native Americans in its collections. Those remains, collected by force as the spoils of colonization, are a reminder of the much larger toll incurred when two long-separated biospheres came into contact: "The greatest destruction of lives in human history," according to the geographer W. George Lovell.

Today, there are a handful of planetary protection officers in the world: the European Space Agency, ESA, has one, as does JAXA, the Japanese space agency. The concept of planetary quarantine, however, is American in origin, and it arose at least partially in response to the catastrophic impact of that initial encounter.

It is impossible to know how many people lived in the Americas in 1491, before European explorers made first contact, but historians estimate that some nine out of every ten people in the New World died in the century or so that followed—most from infectious diseases. In part because there were almost no animals suitable for domestication in the Americas—and thus far fewer opportunities for zoonoses to jump species into humans—the transfer of pathogens in the Columbian exchange was all one way. (Syphilis, which was historically considered to be American in origin, is now thought to have likely been present, albeit in a slightly different form, in pre-Columbian Europe, Asia, and Africa.) Before conquistadores had even set foot in the major cities of South and Central America—Cuzco, in what is now Peru, and Teotihuacán, in Mexico—their microbes had traveled ahead of them, passed from body to body, causing mass deaths, followed by famine and social breakdown.

Without any previous exposure to smallpox, measles, influenza, typhus, and diphtheria, the indigenous people of the Americas had no immunity to these common Old World diseases—and no concept of quarantine, having never had much need for it. The resulting epidemics were surprising only in the sheer scale of their devastation and horror. As Nahuatl testimony transcribed by a Spanish friar, Bernardino de Sahagún, recounts, in a melancholy but clinical tone, "the pustules that covered people caused great desolation; very many people died of them, and many just starved to death; starvation reigned, and no one took care of others any longer."

In 1957, as the Soviets successfully launched Sputnik, and the Cold War militarization of space began to ramp up, some scientists began to worry that the encounter between Earth organisms and any lifeforms that might exist elsewhere in the solar system might also result in mutually assured destruction. Before NASA was even launched, in 1958, the Stanford microbiologist Joshua Lederberg had begun making the case for an international agreement to prevent the contamination of extraterrestrial environments with Earth life, and vice versa. "We are in a better position than Columbus was to have our cake and eat it too," he wrote, arguing that planetary quarantine was essential to the "orderly, careful, and well-reasoned extension of the cosmic frontier." (Lederberg, as it happens, also wrote a landmark 1992 report on emerging infectious diseases in humans; the CDC's Marty Cetron credits this report with inspiring his own interest in the field.)

Lederberg seems to have been primarily motivated by concern for the scientific loss that would occur if Earth life wiped out alien life, rather than the ethical dimensions of such a tragedy. "The overgrowth of terrestrial bacteria on Mars would destroy an inestimably valuable opportunity of understanding our own living nature," he argued.

Others felt humans had a moral responsibility to avoid causing harm elsewhere in the galaxy. C. S. Lewis, better known for chronicling Narnia, also wrote a space-themed trilogy in which he despaired at the idea that a flawed and sinful humanity, "having now sufficiently corrupted the planet on which it arose," would overcome "the vast astronomical distances which are God's quarantine regulations" and "seed itself over a larger area." In the science community, one of Lederberg's allies, the young astronomer Carl Sagan, later wrote that if there was life on Mars, humans must leave the planet alone. "Mars then belongs to the Martians, even if they are microbes," he declared.

Largely as a result of Lederberg and Sagan's campaign, the International Council of Scientific Unions, a nongovernmental organization dedicated to international cooperation in the advancement of science, formed COSPAR, the Committee on Space Research, which still sets the ground rules for extraterrestrial exploration today. Bringing Soviet and American researchers into agreement during the Cold War was not easy, particularly when the space race was so entangled with military dominance. COSPAR ended up settling on the Lederbergian position of protecting the science we might want to do on other planets, as opposed to the planets themselves.

Cassie Conley, a small, quirky woman who often wears her hair in a long hippie braid, is personally more aligned with Lewis. "I don't particularly like humans," she told us, as we sat in her office, the light outside fading. "I think we screwed up this planet well enough that we don't deserve another one—but that's just my personal bias and I'm very careful not to bring it into my job."

Conley got that job when some of the tiny worms she had sent into orbit aboard *Columbia*, in order to study muscle atrophy in microgravity, were found to have survived the space shuttle's disastrous explosion. Her experiment provided an inadvertent demonstration that multicellular life might be able to survive a meteor impact—and thus potentially spread between planets on meteors—and it caught the eye of then Planetary Protection Officer John Rummel. Rummel invited Conley to Washington on a year's placement, then, as he gradually eased his way toward the exit, left her to inherit the role of planetary police officer. (On her crowded bookshelves, between a battered copy of *UNIX for Dummies* and stacks of *Astrobiology Magazine*, she had a sheriff's badge that read "Conley, Planetary Protection Officer, 007." Conley was technically the sixth person in the job, but Rummel held the position twice, so it's a reasonable fudge. "I was trying to get some humor out of a job that's rather challenging at times," she explained.)

As a scientist, Conley is deeply curious about what we might find elsewhere in the universe. "I'm very interested in understanding the evolution of life," she told us. But she is more invested in ensuring that we don't do something that precludes the possibility of answering those questions before we even have the ability to ask them. "The best way to prevent forward contamination is simple: don't go there," she said. "But we've already decided we want to go there, so it's a case of: in the absence of information, don't do something that might reduce your ability to get information in the future."

Back in the 1960s, as the scientific community tried to decide what form planetary protection should take, NASA engineers were faced with two irreconcilable demands: internally, management insisted that anything the agency sent into space must be utterly sterile, while, on national television, John F. Kennedy promised that America would put a man—and his trillions of accompanying bacteria—on the Moon by the end of the decade. In the absence of any absolute certainties, COSPAR dithered, eventually deciding that planetary quarantine would have to operate based on a complex algebra of acceptable risk, in which the probability that a viable microbe would be brought to a planet on a lander would be divided by a guesstimate of how likely it was to survive there, in order to arrive at a global contamination allowance that could be divided between each spacefaring nation.

To fill in the parameters in that formula, NASA began looking at the bacterial kill rates of different sterilization techniques used in the food-processing industry, as well as in the army's bioweapon laboratories at Fort Detrick. Using a particularly hardy spore-forming bacteria as their model for a series of tests, NASA scientists fumigated, irradiated, and baked spacecraft components before smashing them to see how many bugs survived, lurking in cracks and in the threads of screws and bolts. They determined that it was possible to clean a spacecraft sufficiently well that only one in every ten thousand landings would transport a viable microorganism—and thus the limits of 1960s sterilization technology became the standard for the first half of the equation.

The likelihood that Earth life could survive on a particular solar system body was even harder to pin down. Given how little scientists knew about conditions elsewhere in the universe at the time, Conley told us that it "was pretty much a case of sticking their fingers in the air and saying 'Hmmm.'" Ultimately, and somewhat arbitrarily, COSPAR recommended that, for planets of biological interest, the total acceptable risk be kept to no more than a one in a thousand chance of seeding another planet with terrestrial life in the course of exploring it. In the end, "acceptable" simply meant a figure that was the best engineers could achieve without breaking the budgets of member states' space agencies.

The total risk—a 0.1 percent chance of contamination—was then divvied up among the spacefaring nations, with the United States, as one of just two spacefaring superpowers, receiving nearly half of the total allocation. Every one of NASA's subsequent planetary missions—the *Viking* probes, *Pathfinder*, and the ill-fated Mars polar lander—has used up some tiny fraction of this actuarial fantasy.

Once astronauts get involved, though, all bets are off. COSPAR's framework is intended to cover only the short window of time during which a planet remains uncontaminated (and thus alien) enough to be of "biological interest." Originally, this period was set at an optimistic twenty years—in the heady days of the space race, scientists estimated that, for example, dozens of missions to Mars would take place during that time, allowing its indigenous biology to be thoroughly understood. It has since been extended.

⟷

The rain was falling in sheets so thick it was hard to see the road in front of our rental car. We had come to Houston, Texas, to visit Building 31 on the campus of NASA's Johnson Space Center. Here, inside a two-story, windowless, bunker-like building, geologic samples of worlds beyond Earth are kept isolated in hermetically sealed containers. Moon rocks brought back by NASA's Apollo program, particles of solar wind retrieved by the *Genesis* mission, and fragments of asteroids returned by robotic missions form an archive of mineralogical conditions as unearthly as it is fascinating.

We ran across the rain-drenched parking lot through an intensifying storm blowing in off the Gulf of Mexico and checked in with NASA's understandably tight security. (In 2002, a Johnson Space Center intern named Thad Roberts stole an invaluable collection of Moon rocks, which he then attempted to sell to an undercover FBI agent; although the rocks were retrieved, they were irreversibly contaminated.)

Clearance obtained, we were met by Judith Allton, a silver-haired, intensely focused woman with a quiet Texas drawl who joined NASA back in 1974. Allton is part of a diverse crew of curators—NASA describes them as "trailblazing women of curation"—working in the agency's Astromaterials Research and Exploration Science Division. They are charged with preserving and protecting some of the rarest objects on—though not *of*—the planet.

In museology, the art of maintaining ancient or fragile artifacts requires extraordinary control over environmental conditions. Temperature, humidity, and exposure to sunlight, among other factors, are all key in determining whether a material from the past can survive the present at all. When those delicate materials have been brought to Earth from alien worlds, the challenge of giving them appropriate conditions for survival becomes all the more daunting; when those same materials might also pose a threat to life on Earth, even a seemingly minor curatorial mishap could end human civilization itself.

In the early 1960s, as NASA scrambled to deliver on JFK's lunar pledge, the agency added "back contamination" to its planetary quarantine concerns. Carl Sagan, who had done so much to ensure that space explorers could avoid so-called forward contamination, or bringing Earth life to other worlds, warned that there was a possibility—albeit exceedingly remote—that returning lunar explorers might be accompanied by hitchhiking germs that could "multiply explosively" on Earth. Indigenous Moon organisms might be innocuous in the hostile environment of their home, scientists worried, but, "when transported to the comparatively lush conditions of the earth," they might reproduce unstoppably, outcompeting Earth life altogether or permanently altering the planet's biosphere through their metabolic processes. "The introduction into the Earth's biosphere of destructive alien organisms could be a disaster," warned a working group on the risk of back contamination. "We can conceive of no more tragically ironic consequence of our search for extraterrestrial life."

By contrast, most of the engineers at NASA thought the very existence of life on the Moon was so unlikely, given the exceptionally harsh conditions of the lunar surface, that it wasn't worth worrying about. As Elbert King, Judy Allton's predecessor as curator of lunar rocks, argued at the time, "If you really wanted to try to design a sterile surface, this was it."

King had, in fact, been one of the first to argue for the necessity of designing a dedicated lunar receiving lab in order to handle geologic samples brought back by astronauts—but his concern was purely to protect the Moon rocks from Earthly contamination, in order to preserve their scientific value. Nonetheless, to an American public primed by a media diet rich with Martian canals, flying saucers, and Orson Welles's infamous radio broadcast of H. G. Wells's *War of the Worlds*, the risk that Apollo astronauts might bring back extraterrestrial pathogens seemed real. Consulted by the Space Science Board, the deputy U.S. surgeon general admitted that "quarantine is a crude concept and a crude approach," but, nevertheless, "a necessary first step" to protect Earth from alien life.

NASA was now faced with a unique challenge in the history of biocontainment. Whereas other facilities, such as the laboratories at Plum Island, were extensively engineered to protect the outside world from what was inside, this lunar receiving laboratory needed to accomplish that while also thinking from the outside in, protecting its contents from the exterior world. The brief called for a kind of double quarantine—and an entirely new facility design.

Allton expressed this dilemma in terms of cleanliness

versus containment. The technologies involved in keeping something contained, Allton explained, are not just different from those needed to keep something clean—they are often the exact opposite, relying, for example, on negative, rather than positive, air pressure. James Goddard, who led the CDC at the time, declared that, because NASA had no data to show that life didn't exist on the Moon, the strictest quarantine was justified, "even if it cost $50 million to implement." (That figure equates to nearly $420 million today, adjusted for inflation; the final facility cost a little over $8 million, or more than $65 million today.) Indeed, Goddard insisted that the CDC would refuse entry to the United States to Apollo astronauts if they were not kept biologically isolated from the world at large until a period of quarantine had certified them harmless.

Thus instructed, NASA set about designing a lunar receiving laboratory that was capable of housing the returning Apollo astronauts for the duration of their three-week quarantine, as well as containing—and not contaminating—lunar samples while testing their pathogenicity. In addition, of course, the same facility should allow space for conducting the scientific research that was the point of the entire exercise. Allton told us that, while the CDC assumed ultimate regulatory control, other agencies also exercised jurisdiction. "The Department of Agriculture was one," said Allton. "They were worried about decimating the world's crops. And Fish and Wildlife was involved: they didn't want to kill all the fish in the streams. If that's your job, you envision a whole different facility than the person who wants to make exquisitely accurate measurements on tiny pieces of rock."

While skeptical NASA engineers were racing, without much enthusiasm, to get the laboratory built and certified in time for Apollo 11's return, the agency also had to design a quarantine logistics chain to transfer the astronauts, the spacecraft, and their geologic payload from a Pacific Ocean splashdown to Houston without breaking containment. The solution involved modifying an Airstream trailer into what NASA called its Mobile Quarantine Facility: its wheels had been removed, and filtration and air-handling systems installed to keep it at negative pressure, along with an intercom system, emergency oxygen, tanks to contain wastewater, and a decontamination lock through which food and Moon rocks alike could be passed back and forth. Airstream's own publicity materials boasted of "several unique interior features," among which were a high-tech microwave oven and "a medical examination table in place of the typical credenza."

On July 24, 1969, when the command module containing Neil Armstrong, Buzz Aldrin, and Michael Collins landed in the Pacific, recovery swimmers opened the escape hatch, handed specially designed biological isolation garments to the astronauts, then closed the hatch while they put them on. The recovery crew then helped the astronauts out of the spacecraft and onto a raft, scrubbing them down with an iodine-based disinfectant before they were airlifted to the waiting aircraft carrier, where they walked into the Airstream through a negative-pressure tunnel. President Richard Nixon was waiting on board, with another helicopter on stand-by to fly him off the ship if any leaks had been detected—at which point, contingency plans specified that the entire vessel would become an isolation unit, remaining at sea for an unspecified period of quarantine.

Inside the trailer with the astronauts was a doctor, William Carpentier, and an engineer, John Hirasaki, who was a newlywed and, as he recalled in his NASA oral history, had just finished reading Michael Crichton's *The Andromeda Strain*, a thriller about the outbreak of a deadly extraterrestrial microorganism. "So you can imagine," Hirasaki said, "there were mixed feelings floating around about is this real or is this imagined or what are the possibilities?"

Still, the mood in the metal can was buoyant. The astronauts pressed up against the Airstream's window to greet Nixon while Hirasaki took care of off-loading the lunar rock samples, vacuum-packing the sealed boxes in three additional layers of heavy plastic before putting them in the transfer lock to be sprayed with concentrated Clorox. These were raced back to the Lunar Receiving Lab in Houston to measure their radioactivity before it decayed: Elbert King,

When the Apollo 11 astronauts returned to Earth from the Moon, they spent three weeks in quarantine. Here, they are seen greeting—and laughing with—U.S. president Richard Nixon from inside NASA's Mobile Quarantine Facility aboard the USS *Hornet*. (*Photograph courtesy of NASA*)

who unloaded the canisters, described the Moon rocks as "like lumps of charcoal in the bottom of a backyard barbecue grill."

Meanwhile, the Mobile Quarantine Facility sailed back to Hawaii aboard the retrieval vessel, before being towed on a flatbed truck through streets lined with well-wishers to the air force base. Inside, Hirasaki said it felt like a party, with the astronauts keen to talk about their experience. "They were very enthused and quite excited," he said. "Everybody was very up, as you can well imagine." There was liquor and food, including an eggs Benedict ready-meal that exploded in the newfangled microwave, and hot showers for all. "I mean, it was a happy little home," Collins recalled, years later. "We had gin on board, had steaks," he said. "I could have stayed in there a lot longer."

Three days later, the crew reached Houston and transferred to the Lunar Receiving Lab, which was promptly declared an official quarantine area by the medical officer of Harris County, Texas. The next day, a small item in the *Federal Register* proclaimed a state of quarantine lasting from 0100 hours on July 21 to at least 0100 on August 11, 1969, "to prevent contamination of Earth by extraterrestrial life," an understated admission that Texas was facing the risk, however tiny, of a celestial pandemic. In the event of an actual alien contagion, officials later revealed, the plan was to bury everyone in the laboratory alive under a mountain of dirt and concrete, sacrificing astronauts and NASA scientists alike. Technicians who worked in the lab had signed an agreement stating that their next-of-kin would not claim their bodies in case of their death.

In reality, nothing so dramatic occurred. Over the next fifteen days, boredom set in. The astronauts were subjected to extensive medical tests and debriefing sessions, but, in between, they played Ping-Pong, watched TV, and read.

(Collins recalled getting through Steinbeck's novella *Of Mice and Men.*) Armstrong celebrated his birthday with cake, and, speaking to reporters at the time, said that quarantine had been going "about as well as you can expect." In the lab, they could talk with their families by phone, but, in a nod to the philatelic tradition of disinfected mail, Armstrong posted an envelope that he had brought on the mission and had signed by all the crew. It was stamped "Delayed in Quarantine at Lunar Receiving Laboratory—Houston, Texas," and subsequently sold at auction for tens of thousands of dollars.

Meanwhile, in another, sealed-off area of the lab, researchers sawed up pristine cores of lunar rock to examine under the microscope and test for radiation and gaseous emissions, in search of clues to the Moon's history and origin, as well as hints of the prevailing conditions of the solar system when the Moon was formed. Less scientifically valuable dust and rock chips were pulverized for the biological examinations. Arriving at a protocol that would prove lunar material safe had proved tricky; the few scientists who thought lunar pathogens might exist could not agree on how to find them. In the end, the Lunar Receiving Lab staff was instructed to "challenge" dozens of different indicator species of plants and animals with Moon materials, then document the results.

Working in biosecure glove boxes under vacuum, a crew of technicians fed Moon dust to cockroaches and houseflies, funneled it into oysters through holes in their shells, injected it directly into the abdomens of specially raised germ-free white mice, and added it to the water in which shrimp and minnows were living. The USDA had agreed that NASA could forgo constructing the kind of facilities required to hold large animals under quarantine; if any signs of disease had been detected in the mice or fish, Moon dust was to have been shipped to Plum Island, in order to be tested on livestock there. Fortunately, no long-term ill effects or "replicating agents" were detected, although "considerable fighting" was observed in the pink shrimp early in the test, and the majority of the oysters died. (During the Apollo 12 quarantine assays, NASA reports boasted that "all oysters remained in excellent health.")

Similarly, seeds of thirty-five different plant species, from wheat to cantaloupe, were planted in sterile growth media infused with lunar regolith, and technicians scraped the leaves of grown plants with yet more crushed Moon dust. All seemed to not only survive but thrive, leading researchers to conclude that lunar material might make a useful fertilizer for Earth crops.

The real challenge, it turned out, was the difficulty of working in the glove boxes under vacuum. The pressure differential was so large that the gloves themselves had to be quite stiff and unwieldy, making the delicate work of handling rocks and injecting mice extremely awkward. Within days, a glove tore, exposing two technicians to the rocks: they then had to join the astronauts in quarantine. When a similar breach happened while handling rocks brought back by Apollo 12, several staff who were in the room at the time managed to "evade quarantine by fleeing the area before guards charged with enforcing quarantine rules [could] arrive," according to a National Research Council report.

The same review concluded that, on balance, the Apollo quarantine program "would have to be judged a failure. It greatly complicated sample processing, yet if lunar material had contained lethal microorganisms Earth would have been infected in two places: the Pacific Ocean, and Houston, Texas." Even the astronauts realized that, to a large extent, what they were doing was what we might call "biosecurity theater": performing containment in order to be seen to be following rules that most people at NASA felt were burdensome and unnecessary. When Collins and Aldrin reminisced about their experience to PBS on the fifty-year anniversary of their historic landing, Collins pointed out that the command module was full of lunar dust and that it had freely vented into Earth's atmosphere both during reentry and into the Pacific Ocean once the hatch was opened. "You have to laugh a little bit," said Aldrin, adding that the recovery team had dropped the rags used to disinfect the hatch into the ocean.

In fact, the CDC had already foreseen these breaches and demanded that the module remain sealed until it was hauled onto the aircraft carrier by crane, but NASA decided that this recovery method might endanger the astronauts. They compromised by installing a miniature vacuum cleaner in the module, so that the astronauts could suck up all the dust on their way home. It was not particularly effective: in his oral history, the engineer Randy Stone, who quarantined with the crew of Apollo 12, recalled that "the dust was just unbelievable. I was gray after I unloaded the spacecraft."

After the Apollo 14 mission, a committee including representatives from the CDC, the USDA, and the Department of the Interior resolved that quarantine was no longer necessary for lunar missions. The four identical mobile quarantine Airstream trailers built for Apollo missions 11, 12, 13, and 14 were decommissioned. Three have since ended up in museums—one unit was randomly rediscovered at an Alabama fish hatchery—and one is still at large, having been transferred to the USDA for fieldwork, and its whereabouts subsequently mislaid.

Meanwhile, in Houston, relieved NASA technicians were able to focus simply on protecting the rocks from Earth. Their first move was to transfer these fragments of the Moon from the hard-to-operate Lunar Receiving Laboratory into a new Sample Storage and Processing Laboratory—the facility we visited. "Except for the gas analysis and radiation counting labs, the LRL was abandoned to the biologists and doctors," Judith Allton wrote in an essay reflecting on her quarter century curating astromaterials. An empty ammunition bunker at Brooks Air Force Base in San Antonio served as a backup in case of a Houston-based disaster: 14 percent of the collection was secretly moved there in 1976, under police escort.

Before we could set foot in the Moon-rock archive, we went through the by-now-familiar process of suiting up in hospital-style coveralls, with bootees over our feet to help mitigate both dirt and static electricity. The room was small and extremely bright, gleaming with overhead track lights and filled with steel and glass cases, themselves also illuminated, wired up to electrical boxes mounted on the synthetic-tile floor. On the far side, a burglary-proof vault door, rated against torch and tool attacks, led to a deeper archive, where the stardust and interstellar particles are kept; the vault is so secure that, in the event of a catastrophic flood or hurricane, the collection can be sealed inside, safe from harm.

Sealed glove ports for accessing the rocks had been covered with white protective sleeves that resembled hairnets, giving the room the feel of a particularly clean and well-stocked industrial kitchen. (The resemblance was not superficial, Allton said: NASA uses meat-cutting bandsaws repurposed from the restaurant industry to slice and prepare new extraterrestrial samples.) Inside the rows of cases were rocks—dozens of them—chunks of breccia, vesicular basalt, and molten glass. Some were mounted above tiny trays to catch any stray grains, many were sealed yet again inside Teflon baggies, some were even sealed a third time—inside jars, inside bags, inside stainless-steel cases. Labels on the

Building 31 of the Johnson Space Center in Houston, Texas, holds samples of worlds beyond Earth, including interstellar particles and Moon rocks. (*Photograph by Geoff Manaugh*)

outside of each box clarified which specific mission the minerals came from: AP-16, for example, meant Apollo 16.

There were baseball-size smooth black rocks and granular, translucent rocks and jagged, silvery rocks that gleamed nearly as brightly as the cases that held them. We paused in front of a three-inch taupe chunk, partially coated in a dusting of white crystals. "This rock is special," said Allton, in a hushed tone. "This is the Genesis Rock." It was collected during Apollo 15, whose mission was to explore lunar geology, and, if possible, return with a sample of original crust material so that scientists could determine when the Moon was formed. The astronauts, David Scott and James Irwin, had been carefully instructed to look for anorthosite: light-colored, coarse-grained rocks that are among the oldest found on Earth.

"Irwin was walking along and he just had a spiritual moment on the Moon," Allton told us. "He saw this rock sitting on a little pedestal of dirt, with its huge crystals reflecting the sunlight, and he said, 'That's what we came for.'" That

The lunar rock collection at the Johnson Space Center represents an unusual curatorial challenge: to preserve materials whose original environments are found nowhere else on Earth. (*Photograph by Geoff Manaugh*)

rock turned out to be more than four billion years old, and, although it had been heated and beaten up during later impacts, scientists have since discovered that it contains traces of water, which adds a layer of mystery to the predominant theory of the Moon's formation.

In the 1980s, Allton recalled, Irwin, who had by then been ordained as a Baptist minister, brought his family to the Sample Storage and Processing Laboratory to see the rock. "He felt God had led him there," she said. Allton is herself a member of the nearby Webster Presbyterian Church, which describes itself as the "Church of the Astronauts." Its most famous congregant was Buzz Aldrin, who took Communion on the Moon, drinking wine from a small chalice that is now in a collection at the church. In a short paper presented at a NASA conference, Allton described Communion as a behavior that could "reinforce the homelink" for astronauts, providing a ritual reconnection with Earth. The necessity of such a "homelink"—a meaningful bridge across barriers of separation—is something people in quarantine feel, as well.

The Webster Presbyterian Church building incorporates space themes throughout, including stained-glass windows that depict nebulae and, embedded in the window frame, two pieces of a meteorite recovered in Mexico in the late 1960s. The rest of the meteorite, it turned out, had been used as a test rock for the containment, processing, and testing technologies that were later employed in the Lunar Receiving Laboratory.

⟷

By the mid-1970s, as the Apollo program drew to a close and as Viking 1 and Viking 2, the first Martian landers, sent back data that painted a picture of a much harsher, drier environment than many scientists had hoped or imagined, it began to seem as if the rest of the solar system was lifeless—making the need for quarantine moot.

In 1976, NASA downgraded its planetary quarantine program to the Office of Planetary Protection. By 1984, COSPAR had officially relaxed the standards for spacecraft sterilization and cleanliness, replacing its probabilistic calculations of acceptable risk with a simpler set of rules that categorized solar system destinations according to their potential biological activity. "But," Cassie Conley told us, "it turns out maybe we relaxed too much."

The problem is that the new rules were still based on quite limited knowledge. Deciding what measures would adequately protect an imagined form of life in an unknown landscape from an inadequately defined threat made writing planetary protection standards an exercise in speculative extrapolation.

In the 1990s, NASA embarked on a series of research

programs designed to reduce this uncertainty. The better the data in their models of probable contamination, the more precisely tailored the level of protection could be, saving money while improving the scope of possible science. (In human terms, this is analogous to implementing quarantine based on data from robust testing and contact tracing, so that most normal life and economic activity can continue, as opposed to a complete, indiscriminate lockdown.)

Over the past two decades, a series of missions has begun to fill in some of the gaps in our knowledge of conditions in the solar system, sending back promising news of briny oceans on a Jovian moon, Europa, and abundant sources of molecular energy on Enceladus, one of Saturn's satellites. A series of probes and rovers sent to Mars has returned signs of liquid water and seasonal methane clouds. "Mars continues to surprise us," Conley said. "This is a good problem to have." Even the Moon seems more interesting than it used to, with recent observations confirming the presence of water ice at its poles and in the "cold traps" created by permanently shadowed regions.

Meanwhile, a lot of new research, much of it NASA-funded, has redefined our understanding of the extraordinary capabilities of Earth microbes. In deep caves and deserts, thermal vents at the bottom of the sea, and even cans of irradiated ground meat, researchers have discovered microbes that can survive crushing pressure, blistering heat, and caustic alkalinity, without sunlight, water, or any of life's typical thermodynamic levers. Many of these so-called extremophiles seem well adapted to Martian conditions, particularly beneath that planet's surface. On our way to visit WIPP, we stopped off at the New Mexico home of the speleo-biologist Penelope Boston, who, at the time, was serving on NASA's planetary protection advisory committee. Boston's first real caving experience was in the nearby Lechuguilla

Cave, where she twisted her ankle, popped a rib, acquired an infection that swelled her eye shut, and discovered several novel organisms—microbes whose metabolism, life cycle, and chemical proclivities rendered them almost unrecognizable as biology.

"I really think that it's the subsurface of Mars where the greatest chance of extant life, or even preservation of extinct life, would be found," Boston told us as we sat on her couch, surrounded by space-themed art and memorabilia. The cave organisms that Boston has since spent much of her career studying exist on an entirely different timescale, she explained, reflecting an environment in which they have few or no predators but extremely limited sources of energy. "I think this is a long-term, evolutionary repository for living organisms," Boston said: the subsurface as host of a shadow biosphere that persists over geological time. On Mars, whose environment seems to oscillate between freezing aridity and something that is perhaps more clement, Boston speculates that subterranean life could lie dormant for millennia, reawakening only when conditions improve.

Boston's work has led her to implement planetary protection protocols here on Earth, to avoid introducing surface life into the extreme depths that she explores. NASA's new Planetary Protection Officer, the biogeochemist Lisa Pratt, has discovered the extraordinary capabilities of subterranean life herself: her earlier research included the discovery of slow-growing bacteria living under enormous pressure at the bottom of a gold mine in South Africa, where they subsist solely on the byproducts of radioactive energy.

Unfortunately, the other extreme environment in which many of these extremophiles thrive is the spacecraft assembly room in which we visited *Perseverance*, the rover NASA was about to send to look for life on Mars. With cruel irony but Darwinian logic, NASA's rigorous cleaning and decontamination processes turn out to inadvertently select for the kinds of microorganisms that don't mind high heat, extreme aridity, and low nutrient levels. Dotted around the facility, in between the rover, its heat shield, and its descent stage, were what the microbiologist Kasthuri Venkateswaran called "witness plates"—two-inch-square samples of the materials used to build the spacecraft, whose surfaces he periodically swabs to develop a snapshot of the room's bacterial inhabitants. This information serves as "contamination knowledge," in NASA's terminology: a passenger list of the rover's likely bacterial hitchhikers.

Venkateswaran, who is universally known as Venkat, told us that his inventories of clean-room biodiversity have revealed the mundane and the extraordinary, living side by side. "I don't want to see the headline that dog shit is in the JPL clean room," he warned us, before admitting that, despite all the precautions, microbes that are found almost exclusively in dogs' guts still show up on his witness plates, shed by engineers with pets. Meanwhile, in 2009, Venkat discovered an entirely new genus of extremely salt- and acid-tolerant bacteria on the surface of a spacecraft, which he named *Rummeliibacillus*, after John Rummel, Cassie Conley's predecessor as Planetary Protection Officer. In 2016, researchers came across *Rummeliibacillus* again, this time in soil from Antarctica. Other novel organisms isolated from the clean room have since shown up in a Colorado molybdenum mine and a hydrothermal vent at the bottom of the Indian Ocean.

Venkat's microbial census serves several purposes. He archives them, storing thousands of strains of bacteria in a special freezer in anticipation of a future scenario in which life is discovered in a returned Martian sample, and researchers need to rule out the possibility that we brought it with us. He also uses them as model organisms with which to develop new cleaning and sterilization technologies. "If we're able to knock these hardies, then we will be able to kill the other stuff also," he explained.

Recently, he has started sending some of the toughest candidates up for eighteen-month stints aboard the International Space Station (ISS), to test whether they might be able to survive a lengthy journey under intense UV radiation. One strain of *Bacillus pumilus*, named SAFR-032 (where SAF stands for Spacecraft Assembly Facility), was damaged but not killed by its vacation in the vacuum of space—which, Venkat told us, means it "could potentially survive for millions of years once deposited on the Martian surface." (He is now analyzing the survivors to see whether their unique UV-resistant biochemistry could be adapted for use in sunscreen.)

Some of Earth's extremophiles are now Martians; that much is evident. "We know there's life on Mars already because we sent it there," NASA's former chief scientist John Grunsfeld admitted in 2015. Whether these microbes can emerge from dormancy and grow—whether they, as Venkat put it, are capable of "making the red planet green"—is much less well understood. NASA's research program to reduce uncertainty was designed to accumulate the data necessary to produce a more efficient planetary quarantine program. But, while it has yielded a wealth of new knowledge about both Earth and space, it seems to have raised more questions than it has answered.

"I'd say that we are very frustrated within the planetary and astrobiology communities," Penelope Boston admitted toward the end of our visit. "We can use all these wonderful instruments that we load onto vehicles like *Curiosity* and we can send them there. We can do all this fabulous orbital stuff. But, frankly speaking, as a person with at least one foot in Earth science, until you've got the stuff in your hands—actual physical samples returned from Mars—there is a lot you can't do."

Perseverance, or *Percy*, as NASA has begun to call it, which embarked on its journey to Mars just a few months after our visit, represents the first step toward alleviating that frustration. We marveled at it from behind stanchions as engineers pointed out the UV spectrometer intended to search for trace organic chemistry, and the tiny chunk of Martian meteorite, recovered from Oman and donated by London's Natural History Museum, that is mounted onto the instrument's robotic arm to serve as a calibration target. We scribbled notes about which deodorant technicians are allowed to wear (Mitchum unscented) on special, shiny blue clean-room paper, bonded with polyethylene so it doesn't shed lint and particles—in the context of spacecraft assembly, normal paper is considered a contaminant. But the part we really wanted to see—the carousel of forty-three cigar-size metal tubes that will ultimately hold the Martian rocks that *Perseverance* will drill and cache—wasn't there.

As it turned out, the tubes were in a nearby building, awaiting final sterilization: oven-baking at high temperatures for an extended period in a process that would damage the other instruments on the rover but will successfully eliminate any trace of terrestrial biochemistry. After that, they would be shipped to Cape Canaveral separately, under a vacuum backfilled with inert gas. "They don't get installed until right before we actually find our way out to the top of the rocket itself, because we want to keep them as pristine as possible," David Gruel, the mission assembly, test, and launch operations manager, told us. "They're certainly the cleanest thing we've ever taken to Mars."

⟷

NASA retrieved lunar rocks on its own, but it will take an international effort to obtain chunks of Mars. *Percy* will drill cores from the most promising spots, fill up the tubes with regolith and rock, seal them, and leave them on the Martian surface. To understand how those samples will return to Earth, we visited the European Space Research and Technology Center on the Dutch coast between The Hague and Amsterdam. On a bright blue day in October, we dodged a steady stream of cyclists passing through the entrance gates on their way to work, in order to meet Gerhard Kminek, ESA's first and only Planetary Protection Officer. He greeted us inside a *Hobbit*-like warren of timber-framed rooms constructed in the 1980s, featuring curved walls, domed junctions, and circular stairwells; scale models of European spacecraft were on display here and there like model ships of an earlier era.

Kminek, who leads the containment facility and planetary protection team in the Mars sample-return working group, in addition to serving as chair of COSPAR's planetary protection panel, was first introduced to the problem of *Percy*'s tubes while doing his doctoral research in oceanography in San Diego. "Our lab tested some of the procedures they wanted to use to clean the canisters," he told us. Now, Kminek spends several evenings a month on teleconferences with engineers in California, working out how to get those canisters back to Earth.

The plan, as he outlined it on a conference room whiteboard, calls for NASA to launch a Sample Retrieval Lander mission in the late 2020s, ferrying an ESA rover and a small rocket to the Martian surface. ESA will launch another mission, the Earth Return Orbiter, at the same time.

Upon its arrival on Mars, the ESA rover will fetch the tubes collected by *Percy* and load them into a container on the rocket; the rocket will fire the white sphere into Mars orbit. The orbiter will then, in a feat of interplanetary choreography, intercept this basketball-size moon as it circles the Red Planet and robotically load it into an internal biocontainment system, before heading home to Earth. If all goes to plan, it will smack into the Utah desert at ninety miles an hour, without disintegrating, sometime in the early 2030s. This intricate handoff is necessary to "break the chain," NASA's shorthand for ensuring that nothing that has been exposed to the Martian biosphere comes into contact with Earth's biosphere, until after the canisters are opened and their contents proved safe under rigorous BSL-4 containment.

"This is not that trivial," concluded Kminek. "What showed up in all the studies is that you need to start ten years before the sample is back. Otherwise, you cannot get everything done." Kminek told us that he and his colleagues have learned from the way in which NASA was blindsided by the CDC's quarantine requirements for the Apollo mission. "In the end, the green light to come back came from a combination of regulatory authorities, led by public health," he said. "The same is still true today—even worse, to a certain degree, because any mission bringing back samples from Mars will be an international mission, so you have to coordinate regulatory agencies from different countries."

ESA and NASA have thus already assembled countless experts into working groups and panels to discuss the mission's quarantine requirements in advance: by soliciting input and building consensus now, the space agencies hope to avoid any unfortunate or expensive surprises down the line. Indeed, when we met Eugene Cole, the biocontainment designer at the National Bio and Agro-Defense Facility in Kansas, he told us that NASA had just invited him to share his thoughts on Martian quarantine. (Cole's recommendation—to use the ISS as an off-world lazaretto of sorts—offers a clear safety benefit, but other experts have argued that it would be almost impossible to perform the complex tests necessary to prove that these extraterrestrial rocks are safe using the limited facilities aboard the space station.)

"Some of the issues are really not technical or scientific, and it's sometimes very difficult to resolve them," said Kminek. One group, the International Committee Against Mars Sample Return, which bills itself as "the people's environmental awareness organization about planetary protection," argues that, given humanity's track record of losing spacecraft due to human error, mechanical failure, and accidental impacts, it is sheer hubris to assume that the Mars sample-return mission's containment will be foolproof. The late Carl Sagan continues to loom large in the discussion: as the group's coordinator, the astrobiologist Barry DiGregorio, explained to *New Scientist* magazine, "Sagan told NASA's Jet Propulsion Laboratory that if they were so sure they could pull off a perfect Mars sample return mission, then they should load up anthrax bacteria into their prototype container, launch it into space and return it to Earth. Of course, the JPL people were horrified."

Others, once again, think that planetary quarantine is overkill. A growing contingent of scientists complains that COSPAR's rules are expensive to implement and put obstacles in the way of scientific discovery. Worse, they say, these precautions are pointless: our solar system's planets have likely already seeded one another with life, whether those organisms were carried by inadequately sterilized spacecraft or by meteorites.

In a debate reminiscent of those surrounding the competing explanations for disease in fourteenth-century Venice, planetary scientists disagree as to whether life has arisen independently wherever favorable conditions are found, or whether it has, instead, already been transmitted around the solar system through asteroid exchange or comet impact.

This latter theory, dubbed *lithopanspermia*, makes any attempt to prevent further spread seem futile. Recently, new evidence has emerged to support it. Venkat, for example, has since found *Bacillus pumilus* SAFR-032, the clean-room bacterial strain whose UV resistance might inspire tomorrow's sunscreen, embedded deep in basalt rocks collected in the Sonoran Desert outside Tucson, Arizona. When he subjected those rocks to ballistics tests, the microbes happily survived an acceleration and impact equivalent to a meteor strike. More than one hundred Martian meteorites have been found on Earth, which, some argue, means that living Martian organisms have probably already arrived along with them.

Indeed, according to lithopanspermian thought, it is entirely possible that Earth life is Martian in origin. As the English physicist Paul Davies has written, "The planets are not completely quarantined from each other. Debris splattered into space by comet and asteroid impacts gets distributed around the solar system. Mars and Earth in particular have been trading rocks throughout their history, and it is clear that microbes could hitch a ride and be transported in relative safety from one planet to the other." Some researchers even believe that Earth itself is under a kind of "galactic quarantine," imposed by more advanced alien civilizations who are concerned that, as the French biologist Jean-Pierre Rospars put it, "it would be culturally disruptive for us to learn about them."

Even among scientists who don't think life is an interplanetary contagion, quarantine's inevitable leakiness has led, as it does in human medicine, to a certain cynicism. The USSR, for example, was notoriously secretive about its planetary protection protocols, leading some to suspect that they were not quite as rigorous as the American equivalent. If Soviet missions have already contaminated both Mars and

Venus with terrestrial organisms, why should the United States go to such lengths to prevent the same outcome? As we have seen during COVID-19, if just one person or group fails to follow quarantine protocols, everyone else's efforts at disease containment can be fatally compromised.

What's more, Kminek reminded us, national agencies are no longer the only organizations capable of space travel. "The scene has changed," he told us. "We have now a lot of private companies that, at least, would like to do that—we'll see how many of them will actually make it, but I think it is unavoidable, sooner or later." Indeed, shortly after we spoke to Kminek, SpaceX launched a completely unsterilized red Tesla Roadster into orbit; in 2019, the Israeli *Beresheet* lander crashed into the Moon carrying an undeclared payload of tardigrades—microscopic "water bears" that are among the hardiest animals on Earth.

The rise of new spacefaring organizations combined with increasing frustration among the space science community have led to something of a crisis in planetary protection. NASA has responded in its usual fashion, commissioning reviews and reports, as well as expanding and reorganizing its Office of Planetary Protection under the leadership of Lisa Pratt, who was hired in 2018. (Other applicants for the job included a nine-year-old boy, Jack Davis, whose handwritten cover letter included the following delightful inducement: "I am young, so I can learn to think like an alien.")

The basic principles of planetary protection are codified under international law. According to Article IX of the Outer Space Treaty, ratified by the U.S. Senate in 1967, states are required to conduct their exploration of other worlds "so as to avoid their harmful contamination and also adverse changes in the environment of the Earth resulting from the introduction of extraterrestrial matter." "There is a responsibility—not for NASA, not for any company, but for the country, because the country has signed the Outer Space Treaty," Kminek explained. Two other articles define liability for damage, and the scope of responsibility, which extends beyond institutional missions to cover any private or commercial space activities that are prepared or launched from the country.

"So the responsibility is clearly defined," said Kminek. "Now, the U.S. is, I think, struggling at the moment with how to actually implement the responsibility." NASA, as was made clear during the lunar sample-return planning process, is not a regulatory agency—so who is supposed to monitor and enforce the nation's planetary quarantine rules?

"It's thorny, indeed—a whole thicket of thorniness," Lisa Pratt confirmed when we asked her about this regulatory gap. "Given the plans of a few visionary individuals—we won't mention their names," she said, laughing, "the U.S. needs to figure out what it's going to do about launch approvals for the Moon and Mars." In 2020, Pratt said, NASA joined a working group with fifteen other government agencies and offices, in order to agree on roles and responsibilities within the framework of a new national planetary protection policy.

With the clock ticking on these Mars missions—not to mention the decade it will take to build an international sample-return infrastructure—Pratt has her work cut out for her. As she reflected on the process set in motion by *Percy*'s successful launch, she began to recite a litany of questions for which there are, as yet, no answers. Even if everything goes exactly as planned, she told us, it remains an open question as to who will certify any future receiving facility as fit for operation; who will regulate the opening, handling, and testing of the Martian materials; and who has the power to eventually pronounce them safe. When it comes to more-problematic scenarios, Pratt has even less clarity. "What agencies make the decision to approve the landing?"

she asked. "What agencies get to have the first look after landing and decide whether or not containment has been breached? What agencies decide what to do if there is a ding or an obvious break in the containment—what's going to happen right then, on the spot, to secure the area and clean it up?"

The stakes are higher yet for future space travelers. When we asked Kminek what would happen if an astronaut showed signs of an unexplained sickness during the return journey from Mars, he was silent for several seconds. "Today?" he asked. "This is not covered." During the Apollo program, NASA unilaterally decided that maintaining quarantine was less important than protecting the health and safety of its astronauts. "The consequences might be different for Mars," said Kminek. "So that still needs to be discussed."

Cassie Conley, NASA's previous Planetary Protection Officer, also stressed the necessity of developing a plan to handle such an incident and to do so long before astronauts are sent to a potentially hazardous location. "There's plenty of countries that have really good Earth-to-space rocket capability," she pointed out. Although it would clearly contravene the Outer Space Treaty, Conley said, "I don't think there would be a way to stop some country from preventing those astronauts from coming back to Earth."

⟷

"Officially, every time we go into space, we go into quarantine first," the retired Italian astronaut Paolo Nespoli told us. "I did a total of five official quarantines—one time at Cape Canaveral and four times with the Russians in Baikonur."

The practice of a prelaunch quarantine period can be traced back to Apollo 7, an eleven-day mission intended to test the future lunar command module. All three astronauts came down with severe head colds, and, in microgravity, the congestion proved more debilitating than on Earth. Tempers frayed, culminating in a minor mutiny in which the astronauts defied ground control by refusing to wear their helmets during reentry and landing, so that they could still "pop" their ears and relieve sinus pressure.

Charles Berry, the NASA flight surgeon at the time, immediately instituted a strict quarantine protocol to protect astronauts from any exposure to germs that might make them sick in space, even refusing to let Richard Nixon have a prelaunch dinner with the Apollo 11 astronauts on that basis. "I guess that's as close as I ever came to getting fired in my life," Berry recalled in his NASA oral history. "If they came down with anything, whatever it was, a cough, a sniffle, or anything else, we were going to have to prove that it didn't come from the Moon."

Nespoli didn't pay much attention to quarantine during his spaceflight days: "You do it because it needs to be done." That said, he was intrigued by the difference between the American implementation of quarantine and its Russian equivalent. "Somehow the quarantine in the United States, it's a very hectic time," he said. NASA keeps astronauts busy with technical meetings, training sessions, rigid mealtimes, and treadmill sessions in the tiny exercise room. "You do this, you do that—a lot of things are happening," Nespoli recalled. The astronauts are confined to a single, artificially lit building, in order to shift their circadian clock onto the flight schedule, or MET (Mission Elapsed Time), as it's officially known. "It doesn't matter if it's three a.m. outside. They tell you it's ten o'clock in the morning and you get bright light," he explained. "You get shifted in this way."

"Now, the Russians, to be frank, they don't care." Nespoli laughed. "They have a completely different attitude." In Baikonur, the crew stay in a huge compound, originally built to house the head of the Russian space agency: the astronaut Scott Kelly wrote that it is "affectionately called Saddam's Palace by Americans," thanks to its marble floors, glittering chandeliers, en suite Jacuzzis, and pressed linen tablecloths. "I wouldn't say it's a vacation, because it's not quite like that, but it's much more relaxed," Nespoli told us. He recalled going for walks in the extensive grounds, unwinding with the help of the in-house massage therapist, and enjoying three-course lunches and dinners every day. "You still do the training and the technical stuff," Nespoli said, but the Russians seem to understand that the crew needs to rest and recharge, letting go of Earthly stress before the rigors of space.

Reflecting on quarantine during the early months of the COVID-19 pandemic, Nespoli was struck with renewed force by this prior experience. "Quarantine in Baikonur is really a moment where you just let go of everything," he said. Beyond its obvious value in risk reduction, under the right circumstances quarantine can also offer an emotional and intellectual buffer—a necessary psychological cushion before crossing from one world to another.

Similarly, the extended isolation posed by space travel itself—trapped for months on end in a confined space, connected to friends and family only through video calls—was made bearable by considering the alternative. "I was glad I was inside," Nespoli said, "because if I was outside, I would have been dead. I think you can look at that as like lockdown in a certain way: you are quarantined, but you feel kind of free, because you are safe."

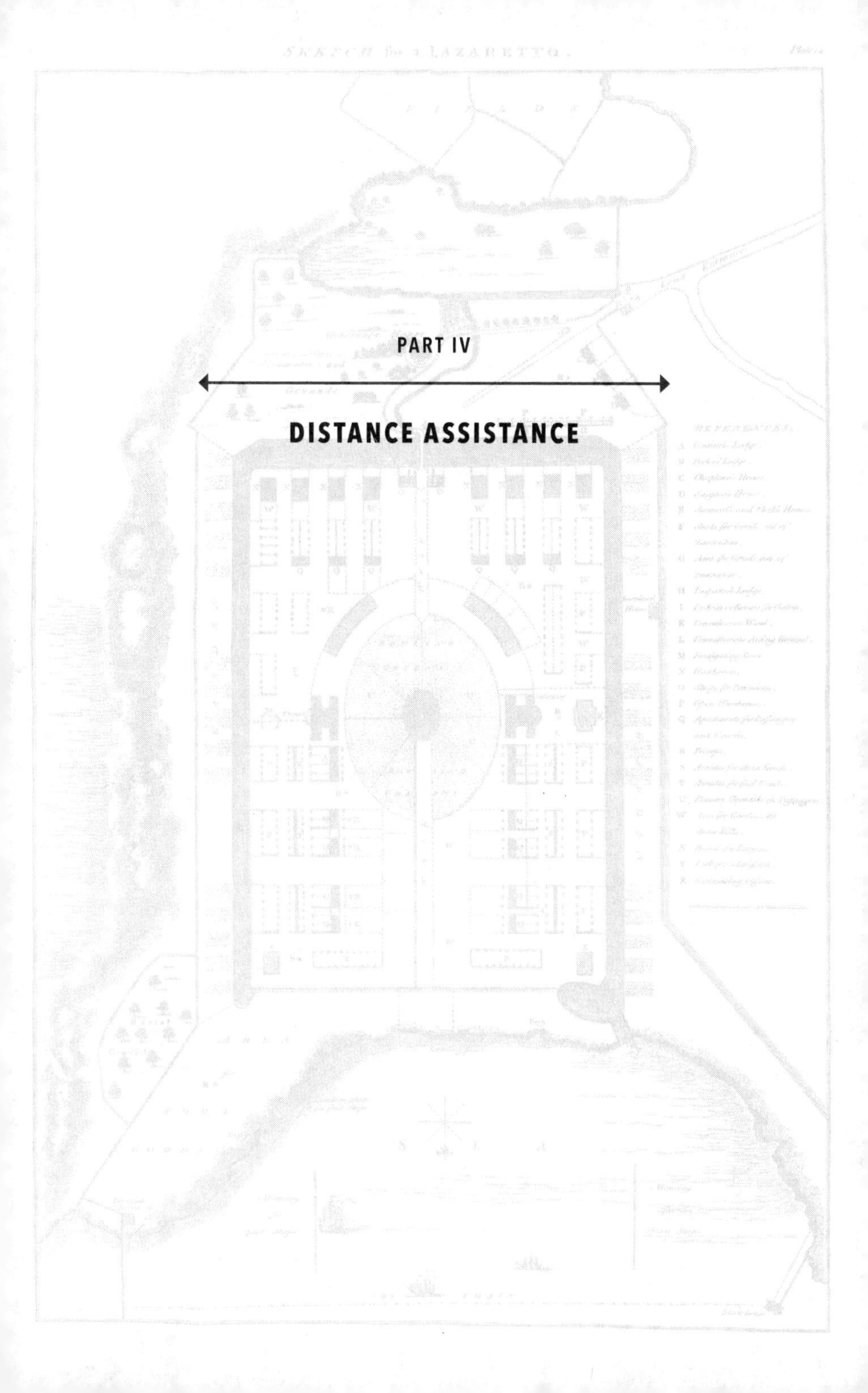

PART IV

DISTANCE ASSISTANCE

9

Algorithms of Quarantine

"We have to think about this on an interstellar scale," said Dr. Kamran Khan. Khan is the founder of BlueDot, a data-modeling and global outbreak-surveillance firm established in 2014. He was referring to the massive data sets, including ticketing information for domestic and international flights, that his company relies on to predict future disease hot spots around the world. Add up all of humanity's combined global air miles, Khan explained, and, every year, we take the equivalent of twenty-two thousand round trips to the sun—interstellar distances, indeed.

We flew to Washington, D.C., during a cold week in March 2018 to attend a pandemic-simulation exercise at the National Academy of Sciences, where Khan—along with the ubiquitous Marty Cetron of the CDC—was speaking. The event was held inside the academy's eccentric neoclassical building less than a mile from the White House; its spacious interior is lined with greening copper wall sconces and mosaic tile patterns, alongside huge doors marked with astrological symbols. Given that plagues and pandemics were once attributed to malevolent influences from the stars, looking at signs of the zodiac while discussing mathematical models of future outbreaks seemed strangely appropriate.

The event was focused on the role played by international air travel in spreading pandemics around the world and the potential role of airports in containing them. Just as steamships and railroads accelerated the movement of people and diseases in the 1800s, and as rockets inspired new fears of alien germs in the 1950s, the dawn of intercontinental flights in the 1920s made aircraft a vector for infection—as the classicist Debbie Felton might put it, air travel opened a new thread of connection, leading previously unknown monsters to our doors. Today, no matter where a disease emerges, public health professionals expect that, after just two or three acts of transmission, it will be detected in one of the world's most connected transportation hubs.

As early as April 1933, an International Sanitary Convention for Aerial Navigation laid down public health requirements for air travel, including guidelines for the design and operation of "sanitary aerodromes," complete with facilities for the isolation of the sick. In December 1944, a Convention on International Civil Aviation followed. Article 14 of the convention's final report specifically obliged sky-faring nations "to take effective measures to prevent the spread by means of air navigation of cholera, typhus (epidemic), smallpox, yellow fever, plague, and such other communicable diseases."

Even the physical design and assembly of airplanes came under scrutiny. In 1953, the air marshal Sir Harold Whittingham, director of medical services at the British Overseas Airways Corporation, the predecessor of today's British Airways, called for engineers to develop "a working knowledge of aircraft construction from a medical point of view." Like Mars rovers assembled at JPL, aircraft would be designed for ease of disinfection, without niches where dangerous germs might grow. (This issue remains a challenge: Walter Gaber, the then medical director of Frankfurt Airport, in Germany, told the conference audience that none of the chemicals certified for medical disinfection are aircraft-component compatible.)

In the 1950s, pesticide-impregnated materials—"DDT in a resin"—were developed for molding things like trays and cabinet surfaces; a London-based firm called Insecta Laboratories devised a microcrystalline "insecticidal coating" for aircraft interiors; and new insect-eliminating, or "disinsection," equipment and procedures were designed to accommodate faster turnaround times for grounded aircraft. The airline industry was then, and is still today, on the front line of disease research and control—think of the deluge of simulations produced in 2020 modeling how coronavirus particles might disseminate through ventilation systems on long-haul flights.

Dozens of public health professionals had come to D.C. that week, joining airport administrators and physicians employed by the largest airlines. The aviation industry faces two fundamentally different fates: airports can either be unwitting hosts of super-spreading events, giving pandemics the space, time, and fresh bodies they need to take hold; or they can, instead, become strategic sanitary choke points, well-managed valves where cases of disease, such as Ebola and SARS, are easily caught and contained. The organizers' hope was that this conference might help to achieve the better of those two options.

The majority of the presentations focused on sharing war stories and practical responses. The emergency management team at Dallas–Fort Worth discussed their Ebola scare, which involved surreptitiously timing a suspect traveler's bathroom visit ("two minutes, thirty seconds—not long enough to vomit"); Frankfurt's Gaber dismissed thermal scanning ("not worth it") before explaining that, during the height of the 2009 swine flu pandemic, Lufthansa began putting doctors on all flights from Mexico as a kind of embedded medical monitoring. Kamran Khan took the conversation in a different direction: the use of mathematical modeling and global network analysis to predict a pathogen's imminent arrival at any given hub.

Khan's goal was to impress upon everyone in attendance that the outlines of the next major pandemic might already be here, lurking inside large data sets—such as airline connections and cargo routes—waiting to be made visible through computation and analysis. BlueDot, Khan's venture capital–backed firm, offers one such method, using proprietary "risk evaluation software" to model when and where a potential outbreak might spread. If dangerous respiratory symptoms are reported by residents of Shanghai, for example, then BlueDot will analyze popular international airline routes departing from that city; it will look at, among other things, the most vulnerable demographic groups living in the most common destinations; then it will run all this against other data sets, such as flight timetables and real-time ticketing information for travelers scheduled to leave Shanghai in the next forty-eight hours. At that point, predictions can begin. "I like to use the word *anticipatory*, rather than *predictive*," Khan cautioned. "We're still in the learning process."

Nonetheless, Khan's data mining has already demonstrated impressive foresight. Khan pointed back to 2015, when BlueDot was part of a team that accurately predicted an outbreak of Zika in Miami, Florida, using airline routes mapped against existing infection hot spots in South America. Nearly two years after this conference in D.C., Khan and BlueDot would beat both the World Health Organization and the CDC in warning about a mysterious new pneumonia-like illness—what we now call COVID-19—emerging in Wuhan, China. (Later, in 2020, Khan and the CDC's Cetron, who are frequent collaborators, would copublish a paper exploring how modeling the passenger base for particular airports—their so-called catchment areas—could help to anticipate the international spread of the coronavirus.)

BlueDot is part of a wave of new private firms and university research institutes looking to combine healthcare data with sophisticated analytic techniques to upend—or, in Silicon Valley terms, to *disrupt*—how we currently think about isolation and quarantine. The approach these groups advocate is pattern-based but precise, even surgical: if a popular air route's duration coincides with a disease's incubation period, and that flight also serves a vulnerable population, then very specific airports—precise gates, terminals, and hotels—can be targeted. Targeting might simply mean increasing passenger exit-screening at a city of origin, or it might mean something more extreme, such as implementing travel bans: the key lies in conducting sufficient surveillance and analysis in order to be able to react appropriately *before* a given disease begins to spread. For individual travelers, of course, this requires a privacy trade-off: your data can be harvested, analyzed, and used to head off future possible pandemics, but this cannot be done without access to your travel plans and, with them, your global location.

In its probabilistic or anticipatory approach to transmission mitigation, BlueDot prioritizes infrastructural connection over static geography; this allows it to estimate when and where these sorts of interventions will be most effective. "It's possible," Khan said, "that if we just think a little bit upstream, maybe the healthcare worker—we can think of a quarantine officer or perhaps a border officer at an airport—might potentially benefit from a heads-up that, hey, in Euclidean space, you're not actually that close to Hong Kong, but through this transportation network, you have a very strong link or connection to it. You might want to be aware of this information."

Dirk Brockmann refers to this prioritization of networked connection over Euclidean space as the study of "effective distance." Brockmann, a physicist and digital epidemiologist based at Humboldt University's Institute for Theoretical Biology in Berlin, has been modeling virtual outbreak scenarios for the past two decades. Like BlueDot, Brockmann uses international flight-route diagrams and an unusual approach to cartography to show that New York and Shanghai, or Paris and Tokyo, although on opposite sides of the planet, are—logistically speaking—much closer than two remote villages in the same country.

In essence, this sort of work is simply an epidemiological update to isochronic geography. To view the world isochronally—a term coined by combining two ancient Greek words, *isos*, or "equal," and *chronos*, or "time"—is to look for locations separated by equal spans of time, not how far apart they are in space. As countless frustrated travelers already intuitively know, for example, a subway ride from New York's John F. Kennedy airport to the Bronx—a distance of roughly sixteen miles—is isochronally longer than an international flight from the same airport to Montreal, nearly four hundred miles away. (A different kind of isochronic milestone was reached in March 2015, when it took less than six hours for a capsule launched from Kazakhstan's Baikonur Cosmodrome to dock with the International Space Station. "It is now quicker to go from Earth to the space station than it is to fly from New York to London," *The New York Times* reported, inadvertently strengthening the case for astronaut quarantine.)

Thinking isochronally forces medical professionals to shift how and where they plan for moments of strategic separation. At its core, quarantine is about managing space and time; as the work of Brockmann and BlueDot, among others, shows, if we change how we think about geography, then the architecture of quarantine must be updated apace.

These same transmission models might even reshape our buildings and cities. "I've actually been contacted by two architects," Brockmann told us. "One of them was concerned about designing an airport terminal—trying to use information from models and interaction patterns to make better designs for airports. And the other was designing a ward in a hospital." The architects' goal was to study computational models of disease transmission, then to design facilities that interrupt those models—that is, to deliberately create spaces in which an infection is less likely to be transmitted person-to-person. We were reminded of a scene in J. G. Ballard's novel *Super-Cannes*: while on a tour of a fictional corporate business park in southern France called Eden-Olympia, the book's protagonist is told that a local doctor wants to institute new health measures. "She's running a new computer model, tracing the spread of nasal viruses across Eden-Olympia," Ballard writes. "She has a hunch that if people moved their chairs a further eighteen inches apart they'd stop the infectious vectors in their tracks."

During the swine flu pandemic of 2009, Brockmann stumbled upon an unusual source of data that he began feeding into his disease models. At the time, an online project called Where's George? was in its eleventh year of tracking the global movement of U.S. currency; the "George" of its title refers to the face of George Washington on the one-dollar bill. That project's voluminous data set, Brockmann told us, coincidentally offered a diagram of human interaction that was of immediate use for a creative epidemiologist looking to understand how people transmit and spread disease. After all, because of its dependence on face-to-face transactions, cash is a superb tracing mechanism for patterns of human contact.

Brockmann had noticed something odd, he told us, in the early mathematical models that he and his colleagues had developed to simulate outbreaks, using these and other data sets. Despite modeling different diseases, with different modes of infection, running different equations, and beginning from different starting points, the resulting networks of transmission took an eerily similar form. Brockmann and a colleague, Dirk Helbing, a professor of computational social science at ETH Zurich, call this the "hidden geometry" of epidemics. In other words, these virtual outbreaks appeared to share a similar—even identical—spreading pattern; the problem was that historical epidemics clearly did not. "It was surprising to me," Brockmann said. "My feeling was, there was something more fundamental going on that we didn't see."

What the models didn't incorporate, Brockmann realized, were response measures: moments of buffering introduced into the system, whether in the form of canceled flights and closed airports or traveler quarantines. "The models treated the epidemic as a spreading phenomenon that goes on unnoticed by the host population," Brockmann said. "But we humans react to things like a pandemic. There are school closures and response strategies. Our models were blind to that." The hidden geometry of disease revealed how we, through tools such as quarantine, can shape our epidemics.

Brockmann discovered a whole new modeling opportunity in this, one that would account for the feedback loops triggered by public health responses. He began looking at what happens to an outbreak model when particular routes of connection are redirected or shut off entirely, to see how this can shape a pandemic—or, ultimately, extinguish one. Brockmann's hunch was that disconnecting just two major nodes—for example, stopping flights from Guinea's

Conakry Gbessia International Airport to Paris Charles De Gaulle, but leaving all other existing routes intact—could offer a precise, cheaper, and much less disruptive way of slowing the global spread of a particular disease, like Ebola. Such measures will have unwanted side effects, from interruptions in local trade to the temporary stranding of foreign aid workers, but these would, in theory, be significantly less intense than those caused by a wider, indiscriminate lockdown. (In terms of Ebola, Brockmann added, history is a dark but influential factor in terms of which European airports confront which epidemiological risks today. Recent Ebola outbreaks have been centered on West Africa and the Congo, Brockmann pointed out, and Germany "didn't have a colonial history in that area": as a result, local and national connections are not as strong, leading to fewer passengers flying to Germany from the hot zone.)

As the COVID-19 pandemic was picking up speed in early 2020, Brockmann joined a team of seven other researchers, all based at Humboldt University, to take these insights and begin generating coronavirus prediction diagrams. The team produced a series of branching maps centered on airports that showed how the coronavirus might spread from one country to the next, sparking new hot spots along the way. "Given an outbreak location and an origin airport close to it," the team wrote, "the model identifies the most probable spreading routes to all other airports in the worldwide air transportation network. Even though passengers can take different routes to a final destination, global spreading patterns are dominated by the most probable paths."

Once these paths have been identified, efforts at quarantine, isolation, and containment can begin—ideally, not after a disease has shown up but in anticipation of its arrival. Through big data and advanced modeling, the logical outcome of this research is that control measures, including quarantine, will soon be anticipatory, not reactive, precision-guided rather than broadly imposed. Brockmann, BlueDot, and dozens of similar projects sketch an almost utopian medical vision: that public health experts will soon be able to stop novel contagions not with the blunt force of travel bans and lockdowns but through surgical incisions in the normal fabric of space and time, with minimal disturbance to the global economy.

Authorities in Dubrovnik and Venice during the Black Death could only have dreamed of such powers: the quarantine grail of maximizing exchange while minimizing risk. This kind of computational rhetoric suggests a coming world where, far from being obsolete or outdated, quarantine will be one of our most effective and pervasive tools for protecting global health—not medieval at all, but futuristic in the literal sense that it will target plagues that have yet to begin.

⟷

In his 2020 book *The Rules of Contagion*, Adam Kucharski examines the hidden rules that govern how diseases—as well as ideas, rumors, and videos—go viral and spread. Kucharski, a biostatistician at the London School of Hygiene & Tropical Medicine, is a member of the U.K. government's influential Scientific Advisory Group for Emergencies (SAGE) subcommittee on modeling. Analysts may promise a future of anticipatory quarantine, but Kucharski's concern is that even the highest-resolution data cannot guarantee certainty.

We met Kucharski for tea and biscuits in the basement cafeteria of his university building, in Bloomsbury, London, a month after the peak of the so-called BBC Pandemic. In what came to be billed as the "largest ever U.K.-wide citizen science project," Kucharski, a team of mathematicians at the University of Cambridge, and advisors at the BBC used cell phone data, contributed by civilian participants who voluntarily downloaded a location-tracking app, to model how a fictional disease—allegedly first detected in Haslemere, a leafy London suburb in the catchment area for Heathrow Airport—might burn through the rest of the U.K. Humans and their cell phones would play the role of cash in Where's George?, illuminating contact networks, and thus transmission patterns, with their daily movements.

For Kucharski, the intensity of an outbreak can affect where, in time, you want to model that particular disease. If health authorities seem to have control over a pathogen—having contained its spread—then, as a modeler, he explained, it is more interesting to reconstruct how specific acts of transmission occurred in the first place. However, if a disease is out of control—a true pandemic—then the recent past is of less interest than the immediate future. "If you've got a flu pandemic with one hundred cases," said Kucharski, "then you're interested in how those one hundred cases came to be. In the early stages, you care a lot about: How did this one transmission event occur? But, when the numbers get big, it's more like: Where's it going next? How long is this going to last? How many beds do we need?"

With scale, in other words, a model's emphasis can shift toward prediction—modeling where future quarantine cordons or pop-up isolation hospitals need to be established—or, as Kucharski carefully clarified, "not *predicting* exactly, but giving people a bit more confidence in making what are very difficult decisions." Even here, though, he said, existing methods often fail.

"There are two approaches to prediction," Kucharski explained. "There's one, which is a more data-driven approach, where you look at trends and follow them on. The other approach is developing a set of mechanisms: you actually simulate an outbreak and assume that there's a population, and that there's transmission going on, and then you see if it matches the data. With the data-driven one, you're not learning anything about the actual process of infection—you're just seeing what the trend is. And with the mechanisms one, you're saying: this is how we think an epidemic should work." At the moment, he told us, both methods perform equally well—which suggests that scientists don't yet understand enough about the rules of contagion to get ahead of an infectious disease. In fact, the idea of orchestrating a virtual outbreak for the BBC came out of the realization that, a century after the Spanish influenza, epidemiologists still did not have enough information from real-world pandemics to validate their assumptions of how disease spreads.

In the team's ensuing review of the BBC Pandemic, the most basic control measures—such as frequent handwashing—were shown to delay the arrival of their simulated contagion in a new town by up to a month. Such interventions quickly flattened the curve, keeping the rate of new infections low enough to avoid hospital bed shortages, which, in turn, helped to prevent healthcare workers from becoming overwhelmed. Based on their analysis, implementing school closures and quarantines would have reduced interactions enough to slow the outbreak still more.

The entire data set from Kucharski's virtual pandemic is now online, the largest of its kind ever made publicly available. Yet even with this boost of granular, location-specific information, the conclusions Kucharski and his colleagues have arrived at can feel unsurprising. The fact that the very young and the very old move the least, for example, or that children have different movement patterns than commuters once you factor in school holidays, is hardly a shock; that models didn't previously account for these details is the real surprise.

As weather forecasters have also found, simply adding more data to a model can yield diminishing returns. "Suppose the earth could be covered with sensors spaced one foot apart," James Gleick writes in his 1987 book, *Chaos*. "Precisely at noon an infinitely powerful computer takes all the data and calculates what will happen at each point." Sadly, Gleick continues, "the computer will still be unable to predict whether Princeton, New Jersey, will have sun or rain on a day one month away."

Kucharski expects the same to be true of pandemic prediction. "It might be that for a flu pandemic, all you can really say is: this is the sort of growth rate we would expect and these are the sort of locations or groups that we think would be most at risk," he told us. "But we couldn't quantify the chance of someone in Birmingham getting infected on this particular day." Models are undeniably useful, to be sure, but they are merely that: models. "In a situation where people have no idea what's going on, even if you can use a model to say, 'Of the options you're considering, don't do that one,' then that's a useful contribution to the evidence base for decisions," Kucharski said.

Three years after our meeting with Kucharski, the exponential spread of COVID-19 presented exactly the kind of data-rich pandemic that he and his fellow epidemiological modelers had, in a sense, been waiting for. "I would say that I'm quite impressed with the advances in the CDC's ability to model, to look at real-time data as close as possible, and to eliminate the lag," Marty Cetron told us, in the autumn of 2020, as a new surge in national case numbers was taking shape. "But I would still say that it's a reactive approach."

One of the ironies of epidemiology, Cetron explained,

is that we are always fighting the *previous* pandemic: it's the one we know best and, thus, it can blind us to what's over the horizon, distracting us with outdated assumptions. Nonetheless, he said, if we want to be ready for the *next* plague, we need to mine the data from this pandemic in order to truly understand what happened, what worked, and why. "I think that's the greatest gift we can give to prepare for managing future pandemics: documentation," Cetron said. "Even if we don't figure it all out in our own day, that needs to be available for people to reflect on and learn from afterwards."

⟷

In October 2019, as a newly emerged coronavirus was likely already causing the first, unrecognized cases of COVID-19 in and around Wuhan, we attended a simulation exercise in which a novel coronavirus sparked a global pandemic, causing sixty-five million deaths in just six months. By that point, inside the simulation, national stock markets had crashed, plummeting by between 20 and 40 percent, with the travel and service industries hit particularly hard. Global GDP was down by 11 percent, triggering a severe, worldwide recession. Large-scale protests and riots had led to the imposition of martial law in some countries; several governments were overthrown.

"It began in healthy-looking pigs," said a news anchor on the fictional GNN network. "A new coronavirus spread silently." Once it jumped species into humans, the disease, called Coronavirus Acute Pulmonary Syndrome, or CAPS, proved to be as lethal as SARS but more transmissible—and, worse, the infected were contagious while still asymptomatic.

For this role-playing exercise, roughly a dozen members of a fictional Pandemic Emergency Board gathered round a U-shaped table in the grand ballroom of the Pierre Hotel, on New York's Upper East Side. They had been selected to provide perspectives from business, public health, and civil society on the pandemic's economic and governance challenges. As the simulated outbreak began to spread beyond its origins in Brazil, several healthcare systems reported shortages of PPE and one country blocked export of the only antiviral drug that seemed effective against CAPS. "What we have is not even enough for our population," said George Gao, the director-general of the Chinese CDC. A senior director at Lufthansa refused to consider halting flights. "We transport infection, yes, but we still must maintain connections," he argued, pointing out that reducing trade and travel could cause more harm than the disease itself.

Three weeks later in the fictional scenario, the group reconvened. Travel bookings were down by nearly half worldwide, and the pandemic had triggered a severe global recession. As central banks rushed to distribute emergency handouts to prop up their collapsing economies, the discussion about which companies were essential—or at least worth bailing out—grew heated. "If you get money out quickly, you have to get accounts in quickly," said Tim Evans, a former senior World Bank official. "We still don't have that capacity, and, without it, you don't have confidence that the money is being well spent."

As global infection rates accelerated exponentially into the winter, the group faced a new challenge: the rise of online misinformation and outlandish conspiracy theories. Some countries, the GNN anchor reported, naming no names, had started to censor online discussion. Gao suggested that training frontline care workers to spread accurate information was the most important strategy to counter an infodemic. Others helpfully pointed out that social trust, community connections, and credibility were essential—but couldn't be built overnight. Evans, the former World Bank official, argued that briefings from the CDC or WHO would seem much more trustworthy if they were "clear about uncertainty," given that the scientific understanding of the disease was still evolving.

Six months in, with no end in sight and no coordinated response on anything from vaccine distribution to messaging, the simulation drew to a close. The global health leaders talked optimistically about the need for advance planning and the potential for public-private partnerships to help tackle the gaps and challenges that had been revealed; the business leaders pointed out that, while collaborating to save both lives and the global economy was a nice idea, they had businesses to run. "The common good is all very well, but we have to go back to our companies on Monday," said Adrian Thomas of Johnson & Johnson.

Event 201—so-named because an average of two hundred epidemic events take place each year, and experts were in agreement that it was only a matter of time before one became a global pandemic—was the fourth major disease outbreak exercise held by the Johns Hopkins Center for Health Security. That it happened to simulate a novel coronavirus pandemic mere weeks before a real one broke out was both pure coincidence and a telling indication as to how likely experts thought that this specific kind of viral respiratory outbreak was to occur.

Johns Hopkins's first such exercise, called Dark Winter, simulated a smallpox attack on Oklahoma. The event was held in June 2001, and its timing, just a few months before 9/11, made its outcome—the near-complete breakdown of government and civil society—deeply resonant. Dark Winter is credited, in part, with spurring George W. Bush to pass Directive 51, a largely classified plan to ensure the continuity of government in the event of a "catastrophic emergency." It was followed by Atlantic Storm in 2005, and Clade X—which we also attended—in 2018. Although the participants and scenarios have varied each time, the outcomes have been equally dire, the lessons depressingly similar.

At Clade X, for example—a domestic governance simulation in which an engineered bioweapon combining the virulence of Nipah virus with the ease of transmission of parainfluenzas, such as bronchitis and pneumonia, had been intentionally released by a fictitious doomsday cult—the lack of leadership was, once again, palpable. Everyone agreed that the president had the final word, but nobody seemed to be responsible for America's outbreak response specifically. The participants returned endlessly to questions about who would brief Congress; who would be capable of authorizing an emergency deployment of military tents as civilian isolation units; who would call state governors to try to ensure a coordinated national response; and who, even, would attend all the funerals. As the former senator Tom Daschle, playing the Senate majority leader, complained, halfway through the day's exercise, "We're five months into this crisis and I still can't tell you who's in charge."

Several challenges were, again, ones that could be solved with more planning and investment. Julie Gerberding, a former director of the CDC, pointed out that global vaccine-manufacturing capacity is insufficient to meet projected demand during a worldwide pandemic. Some of that day's dilemmas revealed vulnerabilities that are hardwired into the American system. Private hospitals, for instance, turned away infectious patients in order to protect the economic interests of their shareholders. (By the end of the simulation, American healthcare had been forcibly nationalized.) Meanwhile, governors enacted state-level quarantines and border cordons.

At the end of each of these events, the organizers published detailed reports and lists of recommendations. Nonetheless, the same problems kept recurring at subsequent tabletop exercises—and, more important, during the first serious pandemic of the twenty-first century. To have simulated a global coronavirus pandemic in October 2019, with the participation of the head of the Chinese CDC, only for COVID-19 to burn through China and then the world months later, was perhaps the most explicit example yet that the most informed models and role-playing exercises can still fall well short of true preparation. The contrast between the promise of anticipatory, precision quarantine as suggested by the work of digital epidemiologists, such as Kamran Khan and Dirk Brockmann, and the inability of global leadership to reach agreement and act decisively even under the simplified conditions of a simulation, was shocking.

"In a way, it's sad that we're still having these conversations," Julie Gerberding told us during a pause at the Clade X event. Back in 2001, she helped lead the CDC's response to that year's postal anthrax attacks. "We have a Department of Defense, we fund it pretty well, and it's pretty stellar," she said. "We still don't have that for this kind of defense."

⟷

For the historian and philosopher Michel Foucault, conditions of pandemic disease present an ideal test lab for trialing new, invasive forms of government control. Foucault, a prominent critic of state power, suggests in his book *Discipline and Punish* that plagues, though disastrous, can be politically convenient. Governments may not respond efficiently to pandemics, or even learn how to do so in the future, but they will likely take advantage of the opportunity to increase "the penetration of regulation into even the smallest details of everyday life."

"The plague," Foucault wrote, "is met by order." In the name of disease prevention and control, he warned, private citizens can be tracked everywhere, supposedly for their own good, required to report their identities, locations, and purposes at any time to meddling authorities. Recall that the first mandatory travel documents, the ancestor of today's passport, emerged during the Black Death in Italy—a useful reminder that government programs enacted during times of quarantine and plague often become permanent features.

Although Foucault's analysis draws on historical events, his remarks are clearly relevant today. The need for cleanliness and sanitation, he writes, can be used to justify public spaces being patrolled and inspected, subject to checkpoints, or cleared of undesirables entirely. As a further barrier against infection, townspeople can be locked inside their own houses—either legislatively, through stay-at-home and quarantine orders, or quite literally, in the case of doors sealed from outside by authorities. During COVID-19, we saw such measures implemented in sites as far-flung as Wuhan and Washington State.

Plagues give states the opportunity to debut—or to reveal previously undisclosed—programs of tracking and confinement that would have seemed ethically unacceptable outside of a public health emergency. For Foucault, pandemics all but inevitably lead to conditions of "permanent, exhaustive, omnipresent surveillance"—what he describes as "a faceless gaze" with "thousands of eyes posted everywhere, mobile attentions ever on the alert." This always-on surveillance, tracking everything, collects what Foucault metaphorically calls "the dust of events, actions, behavior, opinions."

This "dust" is what technology companies today would describe as *metadata*. Indeed, one of the most striking aspects of Foucault's critique, which is focused almost entirely on the perils of government overreach, is that it does not adequately anticipate that the people stepping in to take advantage of pandemics would be corporations, not states; surveillance healthcare and, with it, quarantine have become big business.

A portrait of Foucault hangs in the offices of Palantir, a data-aggregation and modeling firm with close ties to the U.S. defense industry. The irony of seeing Foucault in a context such as this is surely deliberate: to date, Palantir's most widely known contracts have been with the alphabet soup of American intelligence agencies, including the CIA, FBI, and NSA. Depending on one's political point of view, Palantir, with its multibillion-dollar stock valuation, is either proof that sophisticated mathematical modeling is finding its rightful place in the work of federal policymakers, or it is a foreboding sign that there is profit to be made using powerful tools of analysis to track—and someday predict—the minutiae of people's lives. (Of course, the firm's services need not be applied only to surveillance: one of its largest customers is Airbus, which uses Palantir's software to keep track of tens of thousands of airplane parts across the complex aircraft-assembly process.)

In 2020, Palantir saw an opportunity. The company began contracting with the British National Health Service to help manage an onslaught of logistical challenges generated in the wake of COVID-19, although the company's exact role in the deal was not disclosed. Palantir then signed a multimillion-dollar contract with the U.S. Department of Health and Human Services to help manage hospital supply chains, including national ventilator stocks. Mere weeks later, the director of the U.S. Federal Emergency Management Agency requested that all state health officials update Palantir on a daily basis with new hospitalization statistics in an effort to track the pandemic's spread.

Later, the firm became involved in coronavirus vaccine distribution efforts, further applying the logic of industrial tracking to human healthcare. "One of the things that's super interesting about software is it allows you to compress time," Alex Karp, the firm's CEO, said during an online discussion hosted by *The Washington Post*. Karp described a future in which, through accelerated vaccine delivery, the political costs of quarantine and lockdown could be bypassed altogether. "Obviously," he continued, "we do not want to attempt to change the basic way in which we live in order to get pandemics under control." Instead, for Palantir, supercharged data acquisition and analysis promise a technological fix to the problem of emerging pandemics.

Palantir is by no means the only company hoping to merge big data, healthcare, and analytic surveillance. Whether it's BlueDot sifting through international flight details, or Adam Kucharski and the BBC feeding the movements of our cell phones into their outbreak models, the new epidemiology—and, thus, the new quarantine—cannot work without detailed and accurate sources of information.

The success of this approach, however, requires giving up our data in exchange for social stability, economic benefits, and timely access to vaccines, laying bare our lives in the promise that these voluntary acts of exposure will be rewarded with medical protection. This requires trust. "Instead of analyzing people's lives without their knowledge," Kucharski writes in *The Rules of Contagion*, "let them weigh up the benefits and risks. Involve them in the debates; think in terms of permission rather than forgiveness. If social benefits are the aim, make the research a social effort."

During the 1500s in Dubrovnik, health authorities began obliging all travelers to record and report any information they had about outbreaks of the plague in neighboring states. Dubrovnik's own ambassadors and foreign consuls were enlisted as part of these early epidemic-intelligence efforts: civilians, merchants, and officials alike were tasked with providing data to help defend the republic against disease. Gathering and analyzing such information took time and immense human effort, but acting on and redistributing the resulting insights—getting the word out to local merchants and ship captains, warning them of potentially infectious goods or suspected hot zones abroad—posed at least an equal challenge.

By contrast, today's most active disease informants, scouring the world for signs of emerging illness, are not necessarily human at all. Their reporting is voluminous, in both quantity and detail, and instantaneously analyzed by machine-learning algorithms run on cloud-based supercomputers. Foucault's metaphor of dust is an apt one: we now shed data and metadata—or data about data—with nearly everything we do, casting off information about ourselves and our preferences in an endlessly growing file of credit card transactions, online browsing histories, social media posts, smart refrigerator inventories, wearable fitness-tracking reports, internet-connected security-camera footage, virtual-assistant microphone recordings, and more.

In pursuit of convenience, efficiency, and entertainment, we are voluntarily subjecting our own daily lives to the "thousands of eyes posted everywhere, mobile attentions ever on the alert," that Foucault warned his readers about in the 1970s. These devices of capture, measurement, counting, and comparison report the number of steps we take each day or our blood-oxygen levels—pieces of information that, on their own, are not always revealing, but, in aggregate, offer an invasive portrait of our health and well-being, even of our personal contacts, political beliefs, and social activities.

It should come as no surprise that profit-oriented private corporations, employing teams of computationally sophisticated analysts, are finding ever-new ways to use this torrent of data in the service of real-time diagnosis. Predictive, computational, algorithmic, anticipatory: the new descriptors for this emerging front of digitally enabled healthcare are many, and each of those words can be equally applied to isolation and quarantine. The future of quarantine—how and where it will be implemented and enforced—is taking shape, datum by datum, in the ever-watchful objects and internet-connected services all around us.

This trend, of course, did not begin with COVID-19—indeed, it did not take a pandemic to push healthcare diagnostics out of the doctor's office and into our phones and appliances. In 2008, a year before the H1N1 swine flu outbreak, Google began analyzing online queries for evidence that its users might be suffering from symptoms of seasonal flu. They called the project Google Flu Trends. Its underlying assumption was that if you are searching for cold medicine, the opening hours of a nearby pharmacy, or lists of relevant symptoms, then you yourself are likely to be sick. By sifting through those patterns, Google's researchers suggested it might be possible to track flu activity in real time, rather than waiting for hospitals and local health authorities to report case numbers a week or two later.

The idea was provocative, and the initial results seemed promising. However, the data it drew upon ultimately proved to be a mess: someone in California looking up flu symptoms for a sick family member back home in Maine would be noted as a potential case on the West Coast. According to Adam Kucharski, Google Flu Trends, using older search data gathered from previous years, successfully reflected seasonal flu peaks between 2003 and 2008; however, when the novel H1N1 swine flu emerged in spring 2009, Google's models dramatically underestimated the outbreak's size. The computational social scientist David Lazer dismissed the project's algorithm as "part flu detector, part winter detector." This sort of inaccuracy is of little use to local public health providers, and the entire project was discontinued in 2015.

Nonetheless, there is compelling—if not necessarily accurate—information hidden in our online activities. A fall 2020 study discovered that an increase in the number of negative reviews for scented candles posted on Amazon and other online retailers correlated with a spike in COVID-19 infections—one of the symptoms of which is anosmia, or the loss of smell. Pulling this kind of signal out of noise in real time, however, remains an unsolved challenge.

If the problem of Google Flu Trends or, for that matter, Amazon scented-candle reviews is that they extrapolate diagnoses of illness from only circumstantially related data, then direct measurement of the human body itself would seem far more reliable. Smart thermometers—thermometers that report their temperature readings back to the devices' manufacturer—are an example of this new class of home diagnostic tool. If a novel virus begins spreading through the Midwest, for example, and if enough families in the region are using smart thermometers to measure their fevers, then this information can be gathered, processed, and modeled, revealing the outlines of literal hot spots in real time. Human bodies—or, more accurately, their temperatures—would become the flashing warning signs of disease activity on a national scale.

With information comes the possibility for corporate profit. As *The New York Times* reported, one smart-thermometer company used what it called "illness data" to sell targeted advertisements, resulting in ads popping up for "cold medications, disinfectants, toothbrushes and even orange juice" on the company's app. In one widely discussed example, the bleach brand Clorox later took advantage of the data, using it to increase their placement of in-store signage wherever the local fever profile was spiking. (During the COVID-19 pandemic, journalists and politicians alike criticized the CDC for the agency's lack of transparency in sharing case numbers, but, as these examples demonstrate, reporting such information opens it up to corporate monetization—with goals that may not overlap with public health.)

The increasingly ubiquitous microphone-enabled smart-home devices—voice-controlled speakers used to perform internet searches, queue up musical playlists for parties, or change the channel on networked TVs—have also been enlisted as pieces of diagnostic infrastructure. In October 2018, the logistics and home-delivery behemoth Amazon was awarded a patent, based on the company's Echo smart-speaker system, with its built-in virtual voice-assistant Alexa, "for voice-based determination of physical and emotional characteristics of users." The patent drawings submitted by Amazon as part of its twenty-one-page patent documentation depict a woman saying, "Alexa, *cough* I'm hungry *sniffle*"; Alexa responds to the comment first by offering a recipe for chicken soup, then by suggesting an Amazon order of cough drops.

Critics have rightly pointed out that there are serious ethical issues associated with such a service. The patent explicitly goes beyond detecting signs of ill-health: it also listens for indicators of sadness and depression. Serving up ads to customers whose voices—or perhaps out-of-character recent queries—suggest that they are feeling emotionally vulnerable or even suicidal raises obvious questions about corporate morality, medical responsibility, and, of course, user privacy. The patent does not address this. Instead, it cites the added potential of using a customer's Amazon purchase history and recent "number of clicks" to determine the most appropriate commercial intervention. One such option, the patent says, is for Alexa to respond to sad customers by asking, "Are you in the mood for a movie?"

The dream of the smart home is already one of universal sensing: an assisted-living environment available not just to the elderly or ill but to anyone who can afford it, at the touch of a virtual button. Seen one way, these always-on sensors are an encouraging sign that healthcare professionals might have much more precise information available to them for making complex decisions about things like school closures, stay-at-home orders, and individual quarantines—if corporations choose to share it. But, as Adam Kucharski pointed out, it is far from guaranteed that all of this additional data will improve our public health prediction capabilities. It is also not hard to imagine, given the enormous pressure on policy makers to take action to contain a disease, that future quarantine orders and household lockdowns could be implemented based purely on flawed computational inference or faulty algorithmic decision-making—and, worse, that those quarantines and lockdowns could be enforced by our own devices. Gone are the days when Venetian authorities needed to lock homes from the outside or nail heavy boards across front doors: soon, our internet-connected homes might simply lock themselves.

Ubiquitous home-sensing is just one way, the journalist Emily Anthes has written, in which we are "letting our buildings play doctor." In her book *The Great Indoors*, Anthes documents a world where the movements of people through domestic interiors are tracked by Doppler radar; falls and accidents are sensed by in-floor vibration-detection sensors; wireless heart-rate monitors are mounted inside our very walls; and skin-color-sensitive cameras are hidden behind one-way bathroom mirrors, watching for signs of a stroke or the flu. Your house will be actively diagnosing you—sifting through your "dust," in Foucault's terms—and concluding, before you notice anything different, that something is wrong.

The retirement homes and assisted-living facilities that Anthes writes about are, in a sense, simply human versions of Chevron's Barrow Island project, where automated, robotic, and machine-learning technologies attempt to maintain pristine environmental conditions through rigorous quarantine. These efficient, high-tech tools seem impressive when used to screen shipping containers for invasive weeds, but when elderly widows, single parents, and the disabled become subject to radar scans, acoustic analyses, and object-recognition algorithms, such pervasive, automated surveillance and control feels ethically fraught, if not dystopian.

On Barrow Island, Chevron's investment in otherwise outrageously expensive quarantine measures can be rationalized by the enormous profit the company generates from the oil-extraction rights to which this gives them access. The industrial use of this same oil, of course, will lead to future emissions that contribute to the long-term, climate change–induced collapse of the very landscape Chevron has spent so much money trying to preserve. A similar misalignment of short- and long-term goals seems likely in the smart-home environments we see emerging today: the cost of precision analysis and targeted quarantine can be justified because they produce a bonanza of patient data, but a healthier populace is not necessarily the preferred or most profitable outcome for bleach companies and medical-device manufacturers.

The prospect that such technology will someday—probably

soon—be used to assist with determining whom to quarantine, and for how long, seems inevitable. Given the proper medical justification, backed up by supposedly unimpeachable data coming from networked sensors, a fully automated in-house quarantine or stay-at-home order might even fall into the legal gray area encompassed by existing public health powers. Either way, it would not be complicated: an internet-search giant already has access to your browsing history, including indications that you might be suffering from a fever or cough. You may also have been looking for nearby pharmacy locations or discounts on cold medication.

That same internet-search giant's home smart-speaker will have been picking up audio cues that you are coughing. (In 2020, researchers at the Massachusetts Institute of Technology announced the development of a neural net capable of accurately diagnosing COVID-19 in even asymptomatic patients based on the sound of their cough.) If that search giant also owns a subsidiary company that manufactures networked home thermostats, wireless door locks, or internet-connected security cameras, then they will likely also have access to data indicating whether you've been setting your thermostat unusually high or that you have not left your house in days due to a debilitating fever.

The basic medical evidence is all there in the form of a corporate-owned, proprietary data set: you are clearly suffering from whatever given disease is currently circulating in your community. This internet-search giant, acting on state authority to reduce community exposure, does not unlock your front door the next day—or perhaps unlocks your front door only upon evidence, confirmed by a recognized healthcare provider, perhaps one that they themselves own, that you are leaving the house to get tested. If you cannot supply such proof, then you will remain suspected of having potential illness—and your movements correspondingly constrained. The same company can even deliver you chicken soup and play a movie while you languish in quarantine. A not-dissimilar moment occurs in Philip K. Dick's science fiction novel *Ubik*, when the book's protagonist finds himself locked inside his own apartment by a smart door. "The door refused to open," Dick writes. "It said, 'Five cents, please.'" The protagonist—lacking spare change and following a verbal argument with the door—realizes there is an easy way out: he grabs a screwdriver and begins dismantling it.

This is an extreme—and entirely speculative—example, but it's worth noting that Amazon's 2018 patent for detecting signs of illness in the voice commands of its customers specifically mentions that responsive smart-home technologies can do more than order throat sweets. These same technologies, Amazon explains, can also "modify home settings."

Quarantine, in the not-too-distant future, might simply become a new mode programmed into the built environment. Your own home or airport hotel room will become a frontline emergency healthcare technology, temporarily imprisoning you for the public good—and for private profit. The pitfalls are manifold. The outcomes that corporations care about are not necessarily those that matter to society at large—or even those that support public health. The algorithms that power their networked products and services also reinforce existing biases, in both medicine and law enforcement. What's more, the data that underpin those models can be twisted to fit almost any storyline or justify any intervention.

And, of course, the data might simply be wrong: you were searching online for flu remedies, but it was for a loved one or because you were doing research for a novel; you have been sneezing lately, sure, but it is only because you haven't dusted; or you are simply curious what an emerging disease's symptoms might be and, being sensitive to cold, you've been running your heat more than your neighbors have. If your profile appears suspicious, it doesn't matter *why*: you will be quarantined. And your smart home will do it for you.

⟷

As COVID-19 lockdowns spread around the world in 2020, requiring people to stay at home—or, at least, six feet away from one another—the problem of tracking human bodies became central to the enforcement of quarantine. An already-existing trend continued: companies looking to profit from data-aggregation and tracking technology had been trying to move into healthcare for years, and the unique exigencies of quarantine gave them an ideal opportunity to do so. Foucault once remarked that prisons resemble factories, which resemble schools, which resemble hospitals, which resemble prisons, in an endless architectural loop of discipline and punishment. In a time of global pandemics and widespread quarantine, however, our cities and homes might also resemble sports arenas—even before such facilities have been transformed into alternate care wards or *fangcang*.

In the fall of 2015, radio-frequency identification (or RFID) chip technology made its on-field debut in the U.S. National Football League, embedded in players' shoulder pads. Although basic RFID technology was first developed during World War II, it did not become widely used until the 1990s, when MIT researchers linked the serial number on each chip to the internet, creating a networked tracking system. Today, Jill Stelfox, vice president of location solutions at RFID pioneer Zebra Technologies, told us, "Every major car manufacturer in the world, with the exception of two, uses our technology in the creation of a car." Stelfox touted the tags' ability to streamline manufacturing, enabling just-in-time inventory control throughout the supply chain—even guiding the torque wrenches at the ends of mechanical arms as they tighten bolts.

The NFL had already been seeking a way to track its players on-field when, in 2013, Stelfox called. Within two years, by the start of the 2015 football season, stadiums around the country had undergone an almost imperceptible architectural change: their upper decks were encircled with a constellation of small receivers that could record the pings emitted by each of a player's shoulders—twelve pings for every second they are on the field. The quarter-size RFID tags contain a battery-powered chip and an antenna that broadcasts a unique identifying code to the receivers, which, in turn, are connected to a server. There, Zebra's software triangulates the distance between a chip's bleep and the individual receivers all but instantaneously, thereby discerning a player's real-time whereabouts.

The resulting data looks like "a lot of ones and zeros," Cris Collinsworth, an NBC football analyst and a former wide receiver for the Cincinnati Bengals, told us. Cleaned up and graphically visualized, however, it maps game play in an entirely new way. The ability to pinpoint a player's position in real time can be used to calculate exactly how far he has run and how fast, as well as to which players he has been in proximity. "I don't think this chip technology is going to overnight revolutionize the game," Collinsworth admitted. "But I do think that the coach that's going to be the most influential in the NFL in the future is going to be sitting in front of a computer for the better part of the day." Collinsworth's suggestion can easily be transposed to the world of pandemic healthcare: successful quarantine officials of the near future will spend more time tracking potential vectors—ones and zeros—on a computer screen than tracing contacts door-to-door.

To Stelfox's undoubted delight, sports-tracking technology is now rapidly becoming healthcare technology. Even within the NFL, wireless proximity-tracking tools assumed an air of medical necessity during the COVID-19 pandemic, alerting team physicians if players and staff had spent too much time packed close together on the sidelines. The organizers of the 2020 London Marathon also briefly planned to use wireless, Bluetooth-powered, proximity-detection sensors to reduce the threat of in-marathon coronavirus transmission. The idea had been to note those runners who spent more than fifteen minutes in close proximity with other runners in order to ensure that, if any of them later developed symptoms of COVID-19, they could be individually notified and tracked. These plans were discarded not because of any fault with the technology, but because COVID-19 case numbers spiked so dramatically across the U.K. that the marathon was not only delayed; it became a closed, "elite only" event.

Location-tracking tech has since crept off our sports fields and racetracks to insinuate itself in our everyday lives. As a *New York Times* headline declared in November 2020, "The Hot New COVID Tech Is Wearable and Constantly Tracks You." In a bid to avoid a financially debilitating campus shutdown, some American universities began using proximity-tracking tools borrowed from the world of sports to enforce social distancing rules, keeping tabs on their students and staff. (At one school, several thousand students protested, signing a petition demanding that wearing the tracking mechanism—known as a BioButton—be voluntary, not enforced.)

Factory floors offer another vision of how such technology might be adapted to enforce isolation in a time of pandemic disease. In June 2020, Amazon debuted hundreds of "distance assistants": AI-powered workstation screens displaying real-time security camera footage of warehouse employees. If two workers were seen walking or standing too close together, in violation of social distancing rules, red circles would appear around their images on-screen, warning them to back away. "Nothing is more important than the health and well-being of our employees," the company wrote in a blog post, "and we'll continue to innovate to keep them as safe as possible."

This broad turn toward passive tracking technology has been motivated in large part by the sheer difficulty of performing rigorous contact tracing in densely populated, highly mobile environments. The hope that contact tracing could happen automatically—without sending teams of interviewers out with exhaustive questionnaires—has inspired dozens of ambitious efforts all over the world to develop methods of flagging at-risk areas and notifying people before they enter them. In South Korea, commuter rail passengers received smartphone warnings that someone in the town ahead had tested positive for COVID-19. In Shenzhen, China, a health-tracking program within the hugely popular WeChat communication app began mapping "infected neighborhoods" based on existing caseload data; as one Shenzhen resident explained to Reuters, "Seeing the map is a psychological comfort. You can't guarantee there won't be fresh cases, but you can avoid an area that's already hit." Around the world, in an effort to reduce crowding and, thus, opportunities for infection, internet-linked QR codes became a kind of solve-all for allowing certain people onto public transportation or into bars and restaurants, even barring suspect individuals entirely.

Through the use of these sorts of technologies, quarantine and isolation can be detached from a specific location or a dedicated architectural structure and pushed out across the broader landscape: a parallel city can be carved out of the existing one by simply limiting people's access to it, not in space, but in time. Potentially infected people can still go out and shop, in other words—but only between certain hours or only on specified days. Such restrictions already exist in the form of license-plate readers and congestion charge programs, limiting who can drive into a city on particular days of the week; that these same initiatives will soon be given medical justification, in a world increasingly subject to globally transmissible pandemics with no vaccine or cure, seems inevitable.

Real-time infection-mapping and restricted-access technologies promise—or perhaps threaten—to make the whole world into a lazaretto, a virtual quarantine facility defined by regulations that force us to avoid the company of others. Over the six-hundred-year experiment in preemptive detention that is quarantine, the state's capacity for identification and tracking in *time* has become at least as important as its ability to hold someone in *space*. In the coming quarantine, you will be able to go anywhere—but you will be watched, measured, and diagnosed the entire time.

In this pervasive yet invisible twenty-first-century lazaretto, the dominant atmosphere will be one of managed insecurity. You might always be infected—and infectious. You might always be at risk—and a risk to others. At any moment, at the flip of a switch, your world will simply go into quarantine mode.

Epilogue: Until Proven Safe

This book began with an assumption: as messy, flawed, potentially infectious, self-interested beings, quarantine lies at the center of our collective history, as well as our shared future. As we finished writing, the emergence of COVID-19 made part of our case for us. The places that have best contained this new disease, reducing mortality with the minimum of economic and social disruption, were those capable of implementing and managing population-scale quarantines.

While reporting this book in the years preceding COVID-19, our insistence that quarantine still had modern relevance was occasionally met with disbelief. Even among public health officials, the ethical, economic, and social costs of mandatory quarantine combine to make it an unpalatable option. At the end of our John Howard–inspired tour of Mediterranean lazarettos, we added a stop in Geneva to meet Dr. Sylvie Briand, who directs the World Health Organization's global preparedness program for infectious diseases. Briand expressed surprise that we were writing more than a purely historical account of quarantine. Although she admitted its utility in limited cases, Briand warned that quarantine is fraught with moral hazard, including discrimination and unequal application, and that other, less invasive measures often create more benefit—or, at least, less collateral damage.

"We don't recommend quarantine usually," she said, almost apologizing for the fact that we had flown all the way to Geneva to hear this.

Others were more convinced of quarantine's future importance, despite its flaws; those very concerns made its reform all the more urgent in a world in which emerging diseases with pandemic potential are becoming more common, rather than less. Before we flew to Europe, we spoke with Colonel Dr. Matthew Hepburn at DARPA. At the time, Hepburn was program manager of a project called Prometheus, focused on rapid prognostic tests for viral respiratory infections. Part of a suite of initiatives aimed at developing the capability to prevent future pandemics, Prometheus has the goal of predicting whether an individual would become contagious following exposure to a pathogen, before he or she develops symptoms.

Such a tool would seem to remove the uncertainty within which quarantine operates. Nonetheless, Hepburn was emphatic that, even if Prometheus succeeds, quarantine will continue to be an essential tool for the military. In frontline scenarios, he told us, when there is a lag between exposure and diagnosis, and where mysterious outbreaks, inadequate medical infrastructure, and rapid geographic movements all overlap, "oftentimes, quarantine is really the only choice that we have."

Throughout history, one school of thought has advocated for eliminating the conditions under which new infectious diseases can emerge and spread; another has maintained that investments in technology are the key to stopping the next outbreak. Both arguments are correct—to a point—but neither alone will solve the problem of future pandemics. Quarantine will remain indispensable: its circuit-breaker capability reducing rates of infection so that healthcare systems are not overwhelmed and vaccine-resistant viral mutations are less likely to occur.

Although the advent of advanced contagion modeling, location tracking, and data mining offers the promise of refining quarantine, rendering it so minimal and precise as to be almost imperceptible, the use of those tools during COVID-19 demonstrated that, in many ways, effective quarantine has changed remarkably little since its origins during the Black Death. Political attempts to draw on the insights from sophisticated computational models in order to optimize the impact and costs of lockdowns resulted in a cacophony of color-coded tiers, phases, and micro-restrictions whose logic was frustratingly opaque to the majority of citizens. Faced with a set of rules that even those tasked with enforcing appeared not to fully understand, many citizens simply followed their own instincts, with frequently disastrous results. Precision quarantine seemed both unfair and unenforceable.

The future of quarantine, as we've seen, is certainly a question of technology—of testing, tracking, surveillance, containment, and control. It is a question of ventilation systems, plumbing networks, and extreme forms of waste disposal and burial. But it is also a question of civility, of a politics and culture of collaboration that allow for awareness of shared responsibility in the face of an unknown disease. The ability to respond to such uncertainty—and to prioritize the collective good by temporarily separating ourselves—requires cooperation and self-sacrifice, mutual trust and humility, from political leaders and citizens alike.

In the United States, one of the most sobering lessons of the COVID-19 pandemic is that, in today's political climate, something as simple as a mask can be portrayed as so egregious an imposition on individual liberties that, in large swaths of the country, wearing one simply was not tolerated. Genuine, community-centric acts of self-sacrifice—such as not seeing friends or family, or avoiding large gatherings, whether they are at birthday parties or bars—were recast as beyond the pale, somehow at odds with the country's founding spirit. In the cultural politics of the United States today, putting country or community first in the name of disease control is no longer considered a sign of patriotism but of spineless surrender to authoritarian control.

It does not have to be this way. Through quarantine, as well as through basic acts of hygiene and social distancing, we can learn to watch out for one another and, in the process, become good neighbors. The future of quarantine lies not only in technology, then, but also in us: we will never have public health if we do not think of ourselves as a public. Individually, our bodies can do only so much to halt the spread of disease; as social animals, quarantine is part of our collective immune system, an outsourced behavioral response to a viral threat. Like our inbuilt physiological defenses, quarantine is a messy patchwork of partially effective responses; it can overreact, turning on its own host, or fail miserably, letting an infection slip through and take hold. Nonetheless, it offers protection that we badly need, as novel diseases continue to emerge from damaged ecosystems, urban sprawl, and factory farms around the world, traveling toward us by cruise ship and plane.

In the coming decades, we will almost certainly find ourselves more dependent on quarantine, not less. In this future world, glimpses of which have already begun to arrive in the form of SARS, MERS, Ebola, and COVID-19, it is all the more imperative that we find new ways to make separating ourselves from others tolerable, that we invent new forms of isolation and quarantine—public health interventions whose costs and benefits are transparent, borne more equally, and reflect our larger societal values.

It is a bleak vision to describe a world in which we all

must live atomized in the name of disease control, but this need not, in fact, be dystopian. Uncertainty can present an opportunity as well as a threat. As the two of us saw again and again in our travels, quarantine is a strange but powerful generator of creativity and connection, as well as an excellent lens through which to discover previously overlooked communities. There is a long and delightfully strange thread connecting Boccaccio's *Decameron*, with its characters escaping the plague to tell one another stories in quarantine, to the couple running their own radio station while under federal lockdown in Omaha. The laughter of the Apollo astronauts sitting in a modified Airstream during their surreal meeting with President Nixon is a message that connects back to the tens of thousands of postcards and letters sent from remote lazarettos, collected by enthusiasts such as Denis Vandervelde. As this book documents, we already know how to make quarantine better. Now, before the next pandemic, and while the experience of COVID-19 is still fresh in our minds, is the time to do that.

Quarantine can and must be redesigned. As we saw in examples from Dubrovnik to Wuhan, from the Johnson Space Center to the International Cocoa Quarantine Centre, from the Army Corps of Engineers retrofitting Manhattan conference centers to internet-connected smart homes, from WIPP to the Trexler isolation unit, we have the technical and material know-how to remake our spaces of isolation—to take quarantine seriously, from the very beginning, as a spur toward design creativity. If it is not built into a project from the start, quarantine will always seem like an expensive and onerous add-on. As Judy Allton, NASA's curator of Moon rocks, noted in a report outlining mistakes made during the Apollo program, integrating the needs of quarantine early in mission planning would have reduced engineers' resistance to its requirements, making the resulting solutions cheaper and, most important, more effective.

Quarantine can and must be reformed. Kaci Hickox helped point the way with her bill of rights for the quarantined: we need clear guarantees that if public health authorities ask us to temporarily give up our freedom of movement, then they will also assume a duty of care, as well as assuring us due process. We also need confidence that the authorities making these promises will deliver on them. There is no epidemic control without trust, Marty Cetron told us, again and again. "When the trust account is bankrupt, you're screwed when you have to make tough decisions," Cetron said. "And trust can only be built over time by truth in action, not by a lot of empty words in an emergency."

Quarantine can and must be reimagined. It is a lived reality, not simply a public health tool, yet it has barely been planned for in terms of its logistical challenges, let alone its emotional toll. Every step of the quarantine process, whether that quarantine is voluntary or enforced, needs to be reconsidered from an experiential point of view. As Erin Westgate suggested to us, quarantine need not be monolithic—why not build *fangcang* as well as develop at-home quarantine kits, then offer individuals the choice? "We do this with kids all the time: 'Do you want to wear the red pajamas or the green pajamas?'" she said. "To the extent that you can persuade people that some aspect of the choice to quarantine is volitional—that it's in their control—it might make quarantine a lot more politically acceptable, as well as personally meaningful."

Quarantine can and must be reframed. Its shifting regulations and restrictions are never going to be *the* correct solution to a disease—a side-effect-free "cure" for infection transmission. Ironically, if quarantine does work—if the pause it inserts is sufficient to curb the spread of disease—it will almost always be perceived as an overreaction. Instead, quarantine should be understood as a process, a population-scale project that embodies the way in which science works to chip away at uncertainty, testing and measuring in order to edge ever closer to understanding the new and strange. We can never get quarantine absolutely *right*—only less wrong.

Quarantine can and must also be culturally reclaimed as an act of personal responsibility, precisely to help avoid some of the dystopian acts of technological enforcement that otherwise seem imminent. If all quarantines presuppose their own failures, then we need to recognize that these failures are, in fact, rarely technical: they come from willful resistance and a misunderstanding of quarantine's true stakes.

Quarantine exists not just to protect ourselves but to ensure the safety of others—of loved ones and strangers both. In the end, it demands nothing more of us than that we take the appropriate space and time; that we simply pause, before venturing out again, until proven safe.

Authors' Note

To avoid confusion, we use "we" or "us" throughout the book, even for a few interviews and events where only one of us was present.

Sources

In the following pages, we provide a list of sources upon which we drew for each chapter. These are not intended to be exhaustive endnotes, but rather a guide to the interviewees, site visits, and reading material that informed our research.

1. The Coming Quarantine

Our timeline of COVID-19's spread throughout the United States and around the world is based on contemporary news reports, social media, personal experience, and expert interviews we conducted with, among others, Dr. Martin Cetron and Dr. Luigi Bertinato. For this chapter, we also interviewed Angela Munari and Debbie Felton.

The Telegram group we were invited to join is called Doomsday Preppers and Survivalists. As this book goes to press, the group is still active.

The following articles published during the early days of the COVID-19 pandemic either were particularly useful or support our timeline:

"China's Unproven Antiviral Solution: Quarantine of 40 Million," Lisa Du, *Bloomberg* (January 24, 2020). "5 Million People Left Wuhan Before China Quarantined the City to Contain the Coronavirus Outbreak," Ashley Collman, *Business Insider* (January 27, 2020). "Wuhan People Keep Out: Chinese Villages Shun Outsiders as Virus Spreads," Cate Cadell and Sophie Yu, Reuters (January 28, 2020). "Tough Quarantine Measures Have Spread Across China," *The Economist* (January 30, 2020). "Chinese Villages Walled Off Against Outsiders as Coronavirus Toll Mounts," Cindy Chang and Alice Su, *Los Angeles Times* (February 1, 2020). "China, Desperate to Stop Coronavirus, Turns Neighbor Against Neighbor," Paul Mozur, *The New York Times* (February 3, 2020). "How China Built Two Coronavirus Hospitals in Just over a Week," Jessica Wang, Ellie Zhu, and Taylor Umlauf, *The Wall Street Journal* (February 6, 2020). "Coronavirus: Wuhan Community Identifies 'Fever Buildings'

After 40,000 Families Gather for Potluck," *The Straits Times* (February 6, 2020). "China Is Tracking Travelers from Hubei," *The New York Times* (February 13, 2020). "China Quarantines Cash to Sanitize Old Bank Notes from Virus," *Bloomberg* (February 15, 2020). "Fear of Contact Is Boosting China's Robot Delivery Services," Minghe Hu, *Inkstone News* (February 21, 2020). "China Adapts Surveying, Mapping, Delivery Drones to Enforce World's Biggest Quarantine and Contain Coronavirus Outbreak," Yujing Liu, *South China Morning Post* (March 5, 2020).

"Officials Bought This Motel to Hold Coronavirus Patients. Working-Class Neighbors Are Enraged," Richard Read, *Los Angeles Times* (March 5, 2020). Footage of the King County Econo Lodge was tweeted by Cole Miller (@ColeMillerTV) of KOMO News (March 6, 2020). "Motel Converted into Quarantine Site Sparks Controversy," Leila Fadel, NPR (March 8, 2020).

"Leaked Coronavirus Plan to Quarantine 16m Sparks Chaos in Italy," Angela Giuffrida and Lorenzo Tondo, *The Guardian* (March 8, 2020). "Mr. President, Lock Us Up!," Dana Milbank, *The Washington Post* (March 10, 2020). U.S. Congressional Testimony, Dr. Anthony Fauci (March 11, 2020). "'Italy Has Abandoned Us': People Are Being Trapped at Home with Their Loved Ones' Bodies amid Coronavirus Lockdown," Antonia Noori Farzan, *The Washington Post* (March 12, 2020). "Landing at Dulles Airport, I Encountered a Case Study in How to Spread a Pandemic," Cheryl Benard, *The Washington Post* (March 15, 2020). "COVID-19 Bulletin #2," Dare County, North Carolina (March 17, 2020). "Privacy Fears as India Hand Stamps Suspected Coronavirus Cases," Roli Srivastava and Anuradha Nagaraj, Reuters (March 20, 2020). "'This Is Not a Film': Italian Mayors Rage at Virus Lockdown Dodgers," Angela Giuffrida, *The Guardian* (March 23, 2020). "Pet Peeves: Animals React to Having Their Humans on Coronavirus Lockdown," Aitor Hernández-Morales, *Politico* (March 23, 2020). "You've Got Mail. Will You Get the Coronavirus?," Nicola Twilley, *The New York Times* (March 24, 2020). "Italian Police Can Now Use Drones to Monitor People's Movements, Aviation Authority Says," Sharon Braithwaite, CNN (March 24, 2020). "Rise of the Drones: Unmanned Vehicles Become Key Tool in Coronavirus Battle," France 24 (March 27, 2020). "Rhode Island Police to Hunt Down New Yorkers Seeking Refuge," Prashant Gopal and Brian K. Sullivan, *Bloomberg* (March 27, 2020). "'Group of Local Vigilantes' Try to Forcibly Quarantine Out-of-Towners, Officials Say," Aimee Ortiz, *The New York Times* (March 29, 2020). "FBI Says Texas Stabbing That Targeted Asian-American Family Was Hate Crime Fueled by Coronavirus Fears," Marc Ramirez, *The Dallas Morning News* (March 31, 2020).

"Coronavirus: Widely Mocked Retreat for Wealthy to Wait Out Pandemic Cancelled," Kari Paul, *The Guardian* (April 1, 2020). In a tweet posted April 5, 2020, United Nations Secretary-General António Guterres (@antonioguterres) warned that "many women under lockdown for

COVID-19 face violence where they should be safest: in their own homes." "A New Covid-19 Crisis: Domestic Abuse Rises Worldwide," Amanda Taub, *The New York Times* (April 6, 2020). "Tennessee Brothers Who Hoarded Hand Sanitizer Settle to Avoid Price-Gouging Fine," Neil Vigdor, *The New York Times* (April 22, 2020). "Why the Warning That Coronavirus Was on the Move in U.S. Cities Came So Late," Lauren Sommer, NPR (April 24, 2020). "Japan's 'Virus Vigilantes' Take On Rule-Breakers and Invaders," Tomohiro Osaki, *The Japan Times* (May 13, 2020). "With 'Kung Flu,' Trump Sparks Backlash over Racist Language—and a Rallying Cry for Supporters," David Nakamura, *The Washington Post* (June 24, 2020). "Party Houses Defying COVID-19 Orders May Have Utilities Shut Off, Mayor Says," Leila Miller, Richard Winton, and Luke Money, *Los Angeles Times* (August 5, 2020).

"Soldiers Around the World Get a New Mission: Enforcing Coronavirus Lockdowns," Kevin Sieff, *The Washington Post* (March 25, 2020). "Brazil Gangs Impose Strict Curfews to Slow Coronavirus Spread," Caio Barretto Briso and Tom Phillips, *The Guardian* (March 25, 2020). On Twitter, the U.S. State Department's Bureau of South and Central Asian Affairs (@State_SCA) congratulated the Taliban for the group's efforts in assisting with containment measures for COVID-19 (April 10, 2020). "Mexican Criminal Groups See Covid-19 Crisis as Opportunity to Gain More Power," Falko Ernst, *The Guardian* (April 20, 2020).

"Half of Humanity in Virus Confinement," Agence France-Presse (April 2, 2020). "Coronavirus: 4.5 Billion People Confined," Agence France-Presse (April 17, 2020).

"'To Be Shut Up': New Evidence for the Development of Quarantine Regulations in Early-Tudor England," Euan C. Roger, *Social History of Medicine* (April 11, 2019).

"Biological Warfare at the 1346 Siege of Caffa," Mark Wheelis, *Emerging Infectious Diseases* 8, no. 2, Centers for Disease Control and Prevention (September 2002). "Diallylthiosulfinate (Allicin), a Volatile Antimicrobial from Garlic (*Allium sativum*), Kills Human Lung Pathogenic Bacteria, Including MDR Strains, as a Vapor," Jana Reiter et al., *Molecules* 22, no. 10 (October 2017). "The Brightest Bulb," Cynthia Graber and Nicola Twilley, *Gastropod* (December 22, 2020).

Although some historians have suggested that formal quarantine restrictions, in fact, originated with a proclamation in Venice in 1374, the argument for Dubrovnik's historical priority is a convincing one. This chapter benefited, in particular, from *Plague Hospitals: Public Health for the City in Early Modern Venice*, Jane L. Stevens Crawshaw (Routledge, 2012); *Contagion: How Commerce Has Spread Disease*, Mark Harrison (Yale University Press, 2012); and *Expelling the Plague: The Health Office and the Implementation of Quarantine in Dubrovnik, 1377–1533*, Zlata Blažina Tomić and Vesna Blažina (McGill-Queen's University Press, 2015).

The World Health Organization's "Annual Review of Diseases Prioritized Under the Research and Development Blueprint," released in February 2018, first mentions Disease X.

The Decameron, Giovanni Boccaccio, trans. Wayne A. Rebhorn (W. W. Norton, 2013). "The Masque of the Red Death," Edgar Allan Poe, public domain (originally published 1842).

"Monster Culture (Seven Theses)," Jeffrey Jerome Cohen, in *Monster Theory: Reading Culture*, ed. Jeffrey Jerome Cohen (University of Minnesota Press, 1996). *On Monsters: An Unnatural History of Our Worst Fears*, Stephen T. Asma (Oxford University Press, 2009). "Monsters and Fear of Highway Travel in Ancient Greece and Rome," Debbie Felton, in *Monster Anthropology: Ethnographic Explorations of Transforming Social Worlds Through Monsters*, ed. Yasmine Musharbash and Geir Henning Presterudstuen (Routledge, 2020).

Our biblical quotations come from the King James Version.

2. The Quarantine Tourist

This chapter is centered on a long trip undertaken around the Mediterranean and Adriatic Seas, into Western Europe, visiting lazarettos, quarantine facilities, research libraries, and hospitals in Croatia, Italy, Malta, England, the Netherlands, and Switzerland. It draws on interviews with Dr. Luigi Bertinato, Ugo Del Corso, Nicolina Farrugia, Gerolamo Fazzini, Dr. Herbert Lenicker, Angela Munari, Snježana Perojevič, Fausto Pugnaloni, Edward Said, and Dr. Anthony Vassallo.

The Works of John Howard, Esq., vol. 1, *Containing the History of Prisons: The State of the Prisons in England and Wales*, John Howard (William Eyres, 1777). *The Works of John Howard, Esq.*, vol. 2, *Containing the History of Lazarettos: An Account of the Principal Lazarettos in Europe Etc.*, John Howard (J. Johnson, D. Dilly, and T. Cadell, 1791). *Memoirs of the Public and Private Life of John Howard, the Philanthropist*, James Baldwin Brown (T. and G. Underwood, 1823). *Memoirs of John Howard, the Prisoner's Friend*, Charles Kittredge True (Hitchcock & Walden, 1878). *The Curious Mr. Howard: Legendary Prison Reformer*, Tessa West (Waterside Press, 2011).

"John Howard on Quarantine," John E. Ransom, *Bulletin of the Institute of the History of Medicine* 6, no. 2 (February 1938). "The Monuments to John Howard at Kherson in the Ukraine," Negley K. Teeters, *The Prison Journal* 29, no. 4 (October 1, 1949). "Ideas and Their Execution: English Prison Reform," Robert Alan Cooper, *Eighteenth-Century Studies* 10, no. 1 (Autumn 1976). "John Howard and Asperger's Syndrome: Psychopathology and Philanthropy," Philip Lucas, *History of Psychiatry* 12, no. 45 (March 1, 2001).

The Yellow Flag: Quarantine and the British Mediterranean World, 1780–1860, Alex Chase-Levenson (Cambridge University Press, 2020). *Plague Hospitals: Public Health for the City in Early Modern Venice*, Jane L.

Stevens Crawshaw (Routledge, 2012). *Florence Under Siege: Surviving Plague in an Early Modern City*, John Henderson (Yale University Press, 2019). *Expelling the Plague: The Health Office and the Implementation of Quarantine in Dubrovnik, 1377–1533*, Zlata Blažina Tomić and Vesna Blažina (McGill-Queen's University Press, 2015).

The Control of Plague in Venice and Northern Italy, 1348–1600, Richard John Palmer (Ph.D. thesis, University of Kent at Canterbury, 1978).

Venezia: Isola del Lazzaretto Nuovo, Gerolamo Fazzini (Ekos Club / ArcheoVenezia, 2004). *Il Lazzaretto di Ancona: Un'opera dimenticata*, Carlo Mezzetti, Giorgio Bucciarelli, and Fausto Pugnaloni (Cassa di Risparmio di Ancona, 1978).

"Farewell to Malta," George Gordon, Lord Byron (1811).

"The Lazaret on Chetney Hill," Peter Froggatt, *Medical History* (January 1964). "Controlling the Geographical Spread of Infectious Disease: Plague in Italy, 1347–1851," Andrew D. Cliff, Matthew R. Smallman-Raynor, and Peta M. Stevens, *Acta Medico-historica Adriatica* 7, no. 2 (2009). "The Renaissance Invention of Quarantine," Jane Stevens Crawshaw, in *The Fifteenth Century XII: Society in an Age of Plague*, ed. Linda Clark and Carole Rawcliffe (Boydell & Brewer, 2013). "Citizenship and Quarantine at Ellis Island and Angel Island: The Seduction of Interruption," Gareth Hoskins, in *Quarantine: Local and Global Histories*, ed. Alison Bashford (Red Globe Press, 2016). "Life in the Quarantine: Lazaretto at Ploče During the Republic," Vesna Miović, in *Lazaretto in Dubrovnik: Beginning of the Quarantine Regulation in Europe*, ed. Ante Milošević (Institute for Restoration of Dubrovnik, 2018).

3. Postmarks from the Edge

This chapter draws on interviews with V. Denis Vandervelde and Alison Bashford. This chapter also benefits from forty-one years' worth of back issues of *Pratique: Quarterly Newsletter of the Disinfected Mail Study Circle*, edited by V. Denis Vandervelde, published from 1974 to 2015. The Disinfected Mail Study Circle can still be found online at disinfectedmail.org.

Disinfected Mail: Historical Review and Tentative Listing of Cachets, Handstamp Markings, Wax Seals, Wafer Seals and Manuscript Certifications Alphabetically Arranged According to Countries, Karl F. Meyer with Carlo Ravasini (Gossip Printery, 1962).

"Contagions and the Mail," Karl F. Meyer, *Clinical Pediatrics* 2, no. 4 (April 1, 1963). "Karl F. Meyer: The Renaissance Immunologist," Charles Richter and John S. Emrich, *The American Association of Immunologists Newsletter* (June/July 2018).

Imperial Hygiene: A Critical History of Colonialism, Nationalism and Public Health, Alison Bashford (Palgrave Macmillan, 2004). *Medicine at the Border: Disease, Globalization and Security, 1850 to the Present*, ed.

Alison Bashford (Palgrave Macmillan, 2007). *Contagion: How Commerce Has Spread Disease*, Mark Harrison (Yale University Press, 2012). *Crisis and Change in the Venetian Economy in the Sixteenth and Seventeenth Centuries*, ed. Brian Pullan (Routledge, 2006). *International Quarantine*, Oleg P. Schepin and Waldemar V. Yermakov (International Universities Press, 1991).

The Crying of Lot 49, Thomas Pynchon (Harper Perennial, 2006).

We learned of the *Melatenwiese* while reading Simon Winder's *Lotharingia: A Personal History of Europe's Lost Country* (Farrar, Straus and Giroux, 2019).

"Global Biopolitics and the History of World Health," Alison Bashford, *History of the Human Sciences* 19, no. 1 (February 1, 2006). "Civilizing the State: Borders, Weak States and International Health in Modern Europe," Patrick Zylberman, in *Medicine at the Border: Disease, Globalization and Security, 1850 to the Present*, ed. Alison Bashford (Palgrave Macmillan, 2007). "Biology at the Border: An Interview with Alison Bashford," Geoff Manaugh and Nicola Twilley, *BLDGBLOG* (October 9, 2009). "The Treaty-Making Revolution of the Nineteenth Century," Edward Keene, *The International History Review* 34, no. 3 (September 2012).

"The History of Quarantine in Britain During the 19th Century," J. C. McDonald, *Bulletin of the History of Medicine* 25, no. 1 (January–February 1951). "Cholera, Quarantine and the English Preventive System, 1850–1895," Anne Hardy, *Medical History* 37, no. 3 (July 1993). "How Did Britain Come to Rule the Waves?," Dan Snow, *BBC History Magazine* (February 2010). "Black Death," Dr. Mike Ibeji, BBC (March 10, 2011). "Leigh-on-Sea Has Been Named the Happiest Place to Live in Britain," Katie Frost, *Prima* (November 29, 2018). "'A Very Unjust Affront'? William Harvey's Experience of Quarantine," Katie Birkwood, Royal College of Physicians blog (March 26, 2020).

"The First Russian Cholera Epidemic: Themes and Opportunities," R. E. McGrew, *Bulletin of the History of Medicine* 36, no. 3 (May–June 1962). "Russia, Cholera Riots of 1830–1831," Yury V. Bosin, *The International Encyclopedia of Revolution and Protest*, ed. Immanuel Ness (Blackwell Publishing, 2009). "What Great Russians Did in Isolation," Georgy Manaev, *Russia Beyond* (March 27, 2020). "The Health Transformation Army," James Meek, *London Review of Books* 42, no. 13 (July 2, 2020).

"Logbook for First Transatlantic Steamship *Savannah*," National Museum of American History (1819). "Crossing the Atlantic," Vaclav Smil, *IEEE Spectrum* (March 28, 2018).

Wall Disease: The Psychological Toll of Living Up Against a Border, Jessica Wapner (The Experiment, 2020).

"Insecure Boundaries: Medical Experts and the Returning Dead on the Southern Habsburg Borderland," Ádám Mézes (master's thesis, Central European University History Department, 2013). "Habsburg Control on the Ottoman Frontier: Medicine, the Military and Vampire Mania in an 18th Century Borderland," William O'Reilly, paper delivered at Cambridge University (May 9, 2017). "Austrian Measures for Prevention and Control of the Plague Epidemic Along the Border with the Ottoman Empire During the 18th Century," B. Bronza, *Scripta Medica* 50, no. 4 (January 2019). "History Matters: Development and Institutional Persistence of the Habsburg Military Frontier in Croatia," Marina Tkalec, *Public Sector Economics* 44, no. 1 (2020).

Dracula, Bram Stoker, public domain (originally published 1897). *The Imperial Archive: Knowledge and the Fantasy of Empire*, Thomas Richards (Verso, 1993).

"The Wadi Halfa Quarantine," B. H. H. Spence, *Journal of the Royal Army Medical Corps* 43, no. 5 (November 1924). "From 'Death Camps' to Cordon Sanitaire: The Development of Sleeping Sickness Policy in the Uele District of the Belgian Congo, 1903–1914," Maryinez Lyons, *The Journal of African History* 26, no. 1 (1985). "Reanalyzing the 1900–1920 Sleeping Sickness Epidemic in Uganda," E. M. Fèvre, P. G. Coleman, S. C. Welburn, and I. Maudlin, *Emerging Infectious Diseases* 10, no. 4 (April 2004).

"The Early History of Yellow Fever," Pedro Nogueira, *Yellow Fever: A Symposium in Commemoration of Carlos Juan Finlay*, Paper 10 (2009). "Yellow Fever in Europe During 19th Century," M. Morillon, B. Mafart, and T. Matton, in *Ecological Aspects of Past Human Settlements in Europe*, ed. P. Bennike, E. B. Bodzsar, and C. Suzanne (Eötvös University Press, 2002). "The 1802 Saint-Domingue Yellow Fever Epidemic and the Louisiana Purchase," John S. Marr and John T. Cathey, *Journal of Public Health Management and Practice* 19, no. 1 (January–February 2013). "Dramatic Effects of Control Measures on Deaths from Yellow Fever in Havana, Cuba, in the Early 1900's," Rita Isabel Lechuga and Ana Cristina Castro, *James Lind Library Bulletin: Commentaries on the History of Treatment Evaluation* (2016).

"How Columbus Sickened the New World: Why Were Native Americans So Vulnerable to the Diseases European Settlers Brought with Them?," David J. Meltzer, *New Scientist* (October 10, 1992).

The Invention of the Passport: Surveillance, Citizenship and the State, John C. Torpey (Cambridge University Press, 2018). "Health Passes, Print and Public Health in Early Modern Europe," Alexandra Bamji, *Social History of Medicine* 32, no. 3 (August 2019). "In Coronavirus Fight, China Gives Citizens a Color Code, with Red Flags," Paul Mozur, Raymond Zhong, and Aaron Krolik, *The New York Times* (March 1, 2020). "Quarantine-Free Pacts Among Low-Risk Areas Offer Hope for Travel," Hardika Singh, *Bloomberg* (June 26, 2020). "Covid Passports Seen as Key to Resuming International Travel," Charlotte Ryan and Tara Patel, *Bloomberg* (November 23, 2020).

A History of the County of Middlesex, Volume 5, *Hendon, Kingsbury, Great Stanmore, Little Stanmore, Edmonton Enfield, Monken Hadley, South Mimms, Tottenham*, ed. T. F. T. Baker and R. B. Pugh (London, 1976).

4. An Extraordinary Power

This chapter draws on interviews with Mark Barnes, Dr. Martin Cetron, James Colgrove, Lawrence O. Gostin, Kaci Hickox, and Krista Maglen. We heard the retired chief justice Charlie LaVerdiere of Maine speak at the National Summit on Pandemic Preparedness, May 22–24, 2019, in Omaha, Nebraska.

Imperial Hygiene: A Critical History of Colonialism, Nationalism and Public Health, Alison Bashford (Palgrave Macmillan, 2004). *The Barbary Plague: The Black Death in Victorian San Francisco*, Marilyn Chase (Random House, 2004). *The Future of the Public's Health in the 21st Century*, Committee on Assuring the Health of the Public in the 21st Century / Board on Health Promotion and Disease Prevention (National Academies Press, 2002). *Quarantine! East European Jewish Immigrants and the New York City Epidemics of 1892*, Howard Markel (Johns Hopkins University Press, 1999). *Women Adrift: Independent Wage Earners in Chicago, 1880–1930*, Joanne J. Meyerowitz (University of Chicago Press, 1991). *The Trials of Nina McCall: Sex, Surveillance, and the Decades-Long Government Plan to Imprison "Promiscuous" Women*, Scott W. Stern (Beacon Press, 2018). *Emerging Infectious Diseases and Society*, Peter Washer (Palgrave Macmillan, 2010).

Ebola Virus Haemorrhagic Fever, ed. S. R. Pattyn (Elsevier / North-Holland Biomedical Press, 1978). *The Hot Zone: A Terrifying True Story*, Richard Preston (Anchor, 1994). "What Does Ebola Actually Do?," Kelly Servick, *Science* (August 13, 2014). "How Ebola Kills You: It's Not the Virus," Michaeleen Doucleff, NPR (August 26, 2014). "Ebola Tissue Tropism and Pathogenesis," Tara Waterman, *Tara's Ebola Site* (Stanford University, 1999). "Ebola's Catastrophic Effect on the Body," Patterson Clark, Darla Cameron, and Sohail Al-Jamea, *The Washington Post* (October 3, 2014). "Doctors Puzzled Why Only Some Ebola Patients Bleed," Laura Geggel, *Live Science* (October 7, 2014). "Ebola Virus: How It Infects People, and How Scientists Are Working to Cure It," Ilana Kelsey, *Science in the News* (Harvard University, October 14, 2014). "Two Drugs Reduce Risk of Death from Ebola," *NIH Research Matters* (December 10, 2019). "Ebola Virus Disease Fact Sheet," World Health Organization (February 10, 2020).

"Ebola-Poe: A Modern-Day Parallel of the Red Death?," Setu K. Vora and Sundaram V. Ramanan, *Emerging Infectious Diseases* 8, no. 12 (December 2002).

"West Africa Ebola Epidemic Is 'Out of Control,'" Sam Jones, *The Guardian* (June 23, 2014). "Ebola and the Fiction of Quarantine," Nicola Twilley and Geoff Manaugh, *The New Yorker* (August 11, 2014). "U.S. Quarantines 'Chilling' Ebola Fight in West Africa: MSF," Jonathan Allen, Reuters (October 30, 2014). "Fear, Hope Mark Life Inside Ebola Center in Sierra Leone: Witness," Benjamin Black, Reuters (December 9, 2014). "Understanding Ebola: The 2014 Epidemic," Jolie Kaner and Sarah

Schaack, *Globalization and Health* 12, no. 53 (2016). "CDC's Response to the 2014–2016 Ebola Epidemic—West Africa and United States," *Morbidity and Mortality Weekly Report* 65, no. 3 (July 8, 2016). "Case Series of Severe Neurologic Sequelae of Ebola Virus Disease During Epidemic, Sierra Leone," Patrick J. Howlett et al., *Emerging Infectious Diseases* 24, no. 8 (August 2018). "World's Second-Deadliest Ebola Outbreak Ends in Democratic Republic of the Congo," Amy Maxmen, *Nature* (June 26, 2020). "Treating the Ebola Outbreak in Western Africa," Elvis Ogweno, *Journal of Emergency Medical Services* (June 1, 2018).

"Survivors of Ebola Face Second 'Disease': Stigma," Boubacar Diallo, *Medical Xpress* (April 27, 2014). "'I Feel I Have No Future': Thousands Orphaned by Ebola Face Stigma," Nina Devries, *Al Jazeera America* (October 15, 2014). "Panic: The Dangerous Epidemic Sweeping an Ebola-Fearing US," Alan Yuhas, *The Guardian* (October 20, 2014).

"Self-Contamination During Doffing of Personal Protective Equipment by Healthcare Workers to Prevent Ebola Transmission," Lorna K. P. Suen et al., *Antimicrobial Resistance and Infection Control* 7, no. 157 (December 22, 2018).

"Development and Use of Mobile Containment Units for the Evaluation and Treatment of Potential Ebola Virus Disease Patients in a United States Hospital," Gregory Sugalski, Tiffany Murano, Adam Fox, and Anthony Rosania, *Academic Emergency Medicine* 22, no. 5 (May 2015).

"Maine School Board Puts Teacher on Leave After She Traveled to Dallas," Matt Byrne, *Portland Press Herald* (October 17, 2014). "Doctor in New York City Is Sick with Ebola," Marc Santora, *The New York Times* (October 23, 2014).

"Her Story: UTA Grad Isolated at New Jersey Hospital in Ebola Quarantine," Kaci Hickox with Dr. Seema Yasmin, *The Dallas Morning News* (October 25, 2014). "Tested Negative for Ebola, Nurse Criticizes Her Quarantine," Anemona Hartocollis and Emma G. Fitzsimmons, *The New York Times* (October 25, 2014). "Infectious Disease Specialist Dr. Anthony Fauci Rejects Mandatory Quarantine," Benjamin Bell, ABC News (October 26, 2014). "Christie's Office: Quarantined Woman Headed to Maine," Sara Fischer, CNN (October 27, 2014). "The Pre and Post Quarantine Life of Kaci Hickox and Ted Wilbur," Paul H. Mills, *Daily Bulldog* (January 25, 2015).

"150 Health Professionals Call for Olympics in Rio to Be Postponed Due to Zika," Rachel Axon, *USA Today* (May 27, 2016). "The Cognitive Bias That Makes Us Panic About Coronavirus," Cass R. Sunstein, *Bloomberg* (February 28, 2020). "What Ebola Taught Susan Rice About the Next Pandemic," Blake Hounshell, *Politico* (August 6, 2020).

"From Bones of Immigrants, Stories of Pain," Jim Dwyer, *The New York Times* (October 13, 2009). "A Quarantine Hospital So Unwelcome That New Yorkers Burned It Down," John Freeman Gill, *The New York Times* (May 8, 2020).

Yick Wo v. Hopkins, 118 U.S. 356 (1886).

"Review: The Rights Revolution and Support Structures for Rights Advocacy," Ann Southworth, *Law & Society Review* 34, no. 4 (2000). "Researcher Documents Gender, Class Bias in Quarantine Law Measures," University of Kansas (August 24, 2015). "Nine Decades of Promiscuity," Nicholas H. Wolfinger, Institute for Family Studies (February 6, 2018). "Deinstitutionalization, Its Causes, Effects, Pros and Cons," Kimberly Amadeo, *The Balance* (September 24, 2020).

"Until Proven Safe: An Interview with Krista Maglen," Geoff Manaugh and Nicola Twilley, *BLDGBLOG* (November 10, 2009).

"Crucial Steps in Combating the Aids Epidemic; Identify All the Carriers," William F. Buckley Jr., *The New York Times* (March 18, 1986). "Fear Him and Fire Him," Charles Krauthammer, *The Washington Post* (June 27, 1986). "AIDS Quarantine in England and the United States," Ronald Elseberry, *Hastings International and Comparative Law Review* 10, no. 1 (Fall 1986). "AIDS: From Social History to Social Policy," Allan M. Brandt, *The Journal of Law, Medicine & Ethics* (December 1, 1986). "Quarantine and the Problem of AIDS," David F. Musto, in *AIDS: The Burdens of History*, ed. Elizabeth Fee and Daniel M. Fox (University of California Press, 1988). "AIDS and the Limits of Control: Public Health Orders, Quarantine, and Recalcitrant Behavior," Ronald Bayer and Amy Fairchild-Carrino, *American Journal of Public Health* 83, no. 10 (October 1993).

"Highly Drug-Resistant TB Spreading Person to Person: Study," Chris Dall, *Center for Infectious Disease Research and Policy News* (January 20, 2017).

"Nonpharmaceutical Interventions Implemented by US Cities During the 1918–1919 Influenza Pandemic," Howard Markel et al., *Journal of the American Medical Association* 298, no. 6 (August 8, 2007). "Doctor Behind 'Flatten the Curve' Urges Bipartisan Response to Outbreak," John Kruzel, *The Hill* (March 20, 2020).

5. Alone Together

This chapter is centered on a visit to the then-unfinished National Quarantine Unit in Omaha, Nebraska. We also toured the Philadelphia Lazaretto, in Tinicum, Pennsylvania, and the High Level Isolation Unit at the Royal Free Hospital in London, United Kingdom. It draws on interviews with David Barnes, Dr. Georges Benjamin, Dr. Martin Cetron, Dr. Ted Cieslak, Dr. Sir Mike Jacobs, Dr. Papy Katabuka, Dr. Patrick LaRochelle, Rachel Lookadoo, (Ret.) Lieutenant General Todd T. Semonite, Dr. Patrick Ucama, and Erin Westgate, as well as a presentation by Matthew Penn at the National Summit on Pandemic Preparedness, May 22–24, 2019, in Omaha, Nebraska.

The Hot Zone: A Terrifying True Story, Richard Preston (Anchor, 1994). "A History of Federal Control of Communicable Diseases: Section 361 of the Public Health Service Act," Katherine L. Vanderhook (Harvard

Law School, 2002). "Pestilence Houses: Immigration in Charleston," National Park Service, Fort Sumter and Fort Moultrie. "On the Other Side of Arrival: An Interview with David Barnes," Geoff Manaugh and Nicola Twilley, *BLDGBLOG* (October 19, 2009).

"Hundreds of Americans Were Evacuated from the Coronavirus Epicenter. Now Comes the Wait," Miriam Jordan and Julie Bosman, *The New York Times* (February 5, 2020). "*Diamond Princess* Repatriation," Centers for Disease Control media statement (February 15, 2020). "How to Prepare for a Coronavirus Quarantine," Michael Brown, KHTS Hometown Station (March 12, 2020).

"First Case of 2019 Novel Coronavirus in the United States," Michelle L. Holshue et al., for the Washington State 2019-nCoV Case Investigation Team, *The New England Journal of Medicine* 182, no. 10 (March 5, 2020). "Evidence for Limited Early Spread of COVID-19 Within the United States, January–February 2020," Dr. Michelle A. Jorden et al., for the CDC COVID-19 Response Team, *Morbidity and Mortality Weekly Report* 69, no. 22 (June 5, 2020).

"Isolating the Sick at Home, Italy Stores Up Family Tragedies," Jason Horowitz and Emma Bubola, *The New York Times* (April 24, 2020). "Covid-19: Breaking the Chain of Household Transmission," Shamil Haroon, Joht Singh Chandan, John Middleton, and Kar Keung Cheng, *British Medical Journal* 370, no. 8258 (August 14, 2020). "Transmission of SARS-COV-2 Infections in Households—Tennessee and Wisconsin, April–September 2020," Carlos G. Grijalva et al., *Morbidity and Mortality Weekly Report* 69, no. 44 (November 6, 2020).

"Sixty Seconds on . . . Nightingales," Abi Rimmer, *British Medical Journal* 368 (March 30, 2020). "'We Drew Up the Plan over a Brew'—Inside Operation Nightingale," Dan Sabbagh, *The Guardian* (March 31, 2020). "Covid-19: Nightingale Hospitals Set to Shut Down After Seeing Few Patients," Michael Day, *British Medical Journal* 369 (May 7, 2020). "Revealed: Government Spent More Than £200m on Nightingale Hospitals," Nick Carding, *Health Service Journal* (June 10, 2020).

Wuhan Diary: Dispatches from a Quarantined City, Fang Fang, trans. Michael Berry (HarperVia, 2020). *Chinese Poetry*, Yan Zhi (People's Literature Publishing House, 2010). *Construction and Operation Manual of Fangcang Shelter Hospital for COVID-19*, Yan Zhi (Global MediXchange for Combating COVID-19, 2020).

"Two Buildings Donated by WHU Alumni Put into Use the Same Day," Xiao Shan (Wuhan University, September 1, 2018). "World's Billionaires List: The Richest in 2020," *Forbes* (March 18, 2020). "Association of Public Health Interventions with the Epidemiology of the COVID-19 Outbreak in Wuhan, China," An Pan et al., *Journal of the American Medical Association* 323, no. 19 (April 10, 2020). "Fangcang Shelter Hospitals: A Novel Concept for Responding to Public Health Emergencies," Simiao Chen et al., *The Lancet* 395 (April 18, 2020).

"Wuhan Coronavirus: From Silent Streets to Packed Pools," BBC News (August 18, 2020). "The Sealed City," Peter Hessler, *The New Yorker* (October 12, 2020). "Large-Scale Public Venues as Medical Emergency Sites in Disasters: Lessons from COVID-19 and the Use of Fangcang Shelter Hospitals in Wuhan, China," D. Fang et al., *BMJ Global Health* 5, no. 6 (2020).

"Isolation or Quarantine: An Interview with Dr. Georges Benjamin," Geoff Manaugh and Nicola Twilley, *BLDGBLOG* (October 27, 2009). "The United States Needs a 'Smart Quarantine' to Stop the Virus Spread Within Families," Harvey V. Fineberg, Jim Yong Kim, and Jordan Shlain, *The New York Times* (April 7, 2020). "Urgent Care," Paige Williams, *The New Yorker* (August 3 and 10, 2020).

"Negative-Pressure Plastic Isolator for Patients with Dangerous Infections," P. C. Trexler, R. T. Demond, and Brandon Evans, *British Medical Journal* (August 27, 1977). "Life in a Germ-Free World: Isolating Life from the Laboratory Animal to the Bubble Boy," Robert G. W. Kirk, *Bulletin of the History of Medicine* 275 (Summer 2012). "'His Life Was His Work': 102-Year-Old Scientist Will Leave Behind Legacy," Virginia Black, *South Bend Tribune* (October 7, 2013). "A History of Isolator and Containment Technology Part 1: Early Containment Leading to Flexible Film Isolators," Doug Thorogood, *Clean Air and Containment Review*, no. 18 (April 2014). "Ebola Virus Disease: The UK Critical Care Perspective," D. Martin et al., *British Journal of Anaesthesia* 116, no. 5 (May 2016).

"In Congo, a New and Less Isolating Ebola Treatment Center," Al-Hadji Kudra Maliro, Associated Press (September 10, 2018). "New Ebola Treatment Centre in DR Congo Lets Patients Speak to Relatives," Associated Press (September 15, 2018). "Congolese Physician Develops Portable Emergency Care Unit Revolutionizing Ebola Care and Treatment," Alliance for International Medical Action (December 12, 2019).

Li Lazaretti della Città, e Riviere di Genova del MDCLVII, Ne quali oltre à successi particolari del Contagio si narrano l'opere virtuose di quelli che sacrisicorno se stessi alla salute del prossimo, E si danno le regole di ben governare un Popolo flagellato dalla Peste, R. P. Antero Maria da S. Bonaventura (P. G. Calenzani, 1658). *A Residence in Greece and Turkey: With Notes of the Journey Through Bulgaria, Servia, Hungary, and the Balkans*, Francis Hervé (Whittaker & Company, 1837). *Little Dorrit*, Charles Dickens, (Modern Library, 2002; originally published 1857). *The Plague*, Albert Camus, trans. Stuart Gilbert (Vintage International, 1991; originally published 1947). "Early Nineteenth-Century Mediterranean Quarantine as a European System," Alexander Chase-Levenson, and "The Places and Spaces of Early Modern Quarantine," Jane Stevens Crawshaw, in *Quarantine: Local and Global Histories*, ed. Alison Bashford (Red Globe Press, 2016). *The Yellow Flag: Quarantine and the British Mediterranean World, 1780–1860*, Alex Chase-Levenson (Cambridge University Press, 2020).

"Love in a Lazzaret," Henry T. Tuckerman, *The Knickerbocker; or, New York Monthly Magazine* 12, no. 6 (December 1838). "The Nineteenth-Century Quarantine Narrative," Kelly Bezio, *Literature and Medicine* 31, no. 1 (Spring 2013). "Budget Beer and Spiked Seltzer Dominated During the Pandemic," Jordan Valinsky, CNN Business (June 10, 2020). "From Neighbourly Romances to Zoom Sex: The Boom in Lockdown Erotica," Pearse Anderson, *The Guardian* (July 6, 2020). "Britons Watched TV for 40% of Waking Hours During Covid Lockdown," Mark Sweney, *The Guardian* (August 4, 2020). "The Flight Goes Nowhere. And It's Sold Out," Tariro Mzezewa, *The New York Times* (September 19, 2020).

"Legal Issues Surrounding Public Health Emergencies," David Fidler, *Public Health Reports* 116, S2 (2001).

6. Biology at the Border

This chapter draws on interviews with Eugene Cole, Stephanie Dahl, Matthew Fulks, Paul Hadley, John Henneman, Stephen Higgs, Michele Jacobsen, Yue Jin, Shahryar Kianian, James Kolmer, Heather Lake, Simon McKirdy, (Ret.) General Richard B. Myers, Noa Pinter-Wollman, Sara Redstone, Matthew Rouse, Jim Stack, Les Szabo, Ron Trewyn, Marty Vanier, and Mitch Vega. It was informed by visits to Kansas State University and the future site of the National Bio and Agro-Defense Facility, both in Manhattan, Kansas; to the Needles Border Protection Station, in Needles, California; to the International Cocoa Quarantine Centre in Reading, U.K.; and to the USDA's Cereal Disease Lab in St. Paul, Minnesota.

An essential source for this chapter was *The Handbook of Plant Biosecurity: Principles and Practices for the Identification, Containment and Control of Organisms That Threaten Agriculture and the Environment Globally*, ed. Gordon Gordh and Simon McKirdy (Springer, 2014).

Biological Pollution: An Emerging Global Menace, ed. Kerry O. Britton (American Phytopathological Society, 2004). *Out of Eden: An Odyssey of Ecological Invasion*, Alan Burdick (Farrar, Straus and Giroux, 2005). *The Death of Grass*, John Christopher (Penguin Classics, 2009). *The Viking in the Wheat Field: A Scientist's Struggle to Preserve the World's Harvest*, Susan Dworkin (Walker Books, 2009). *American Chestnut: The Life, Death, and Rebirth of a Perfect Tree*, Susan Freinkel (University of California Press, 2009). *California Agriculture: Dimensions and Issues*, ed. Philip L. Martin, Rachael E. Goodhue, and Brian D. Wright (University of California, Giannini Foundation of Agricultural Economics, Division of Agriculture and Natural Resources, 2003). *Big Chicken: The Incredible Story of How Antibiotics Created Modern Agriculture and Changed the Way the World Eats*, Maryn McKenna (National Geographic, 2017). *Wheat Rusts: An Atlas of Resistance Genes*, R. A. McIntosh, C. R. Wellings, and R. F. Park (CSIRO Publications, 1995). *Fasti*, Ovid, trans. A. J. Boyle and R. D. Woodard (Penguin Classics, 2000). *Spillover: Animal Infections and the Next Human Pandemic*, David Quammen (W. W. Norton, 2012).

"Social Spiders," Duncan E. Jackson, *Current Biology* 17, no. 16 (August 21, 2007). "Ant Traffic Rules," Vincent Fourcassié, Audrey Dussutour, and Jean-Louis Deneubourg, *Journal of Experimental Biology* 213, no. 14 (July 2010). "Want to Get Out Alive? Follow the Ants," Conor Myhrvold, *Nautilus* (May 8, 2014). "Sacrificial Virgin Spiders Let Their Nieces Eat Them Alive," Sandhya Sekar, *New Scientist* (September 18, 2017). "Associate Professor Builds Connections Between Ant Nests, Human Architecture," Teddy Rosenbluth, *The Daily Bruin* (August 6, 2018). "The Impact of the Built Environment on Health Behaviours and Disease Transmission in Social Systems," Noa Pinter-Wollman et al., and "Interdisciplinary Approaches for Uncovering the Impacts of Architecture on Collective Behavior," Noa Pinter-Wollman et al., *Philosophical Transactions of the Royal Society B: Biological Sciences* 373, no. 1753 (August 19, 2018). "Social Interactions Shape Individual and Collective Personality in Social Spiders," Edmund R. Hunt et al., *Proceedings of the Royal Society B: Biological Sciences* 285, no. 1886 (September 12, 2018). "Social Network Plasticity Decreases Disease Transmission in a Eusocial Insect," Nathalie Stroeymeyt et al., *Science* 372, no. 6417 (November 23, 2018). "Sickness Effects on Social Interactions Depend on the Type of Behaviour and Relationship," Sebastian Stockmaier et al., *Journal of Animal Ecology* 89, no. 6 (June 2020). "In Social Insects, Researchers Find Hints for Controlling Disease," Michael Schulson, *Undark* (July 22, 2020). "Emerging Infectious Disease and the Challenges of Social Distancing in Human and Non-human Animals," Andrea K. Townsend et al., *Proceedings of the Royal Society B: Biological Sciences* 287, no. 1932 (August 12, 2020).

"Origins of Major Human Infectious Diseases," Nathan D. Wolfe, Claire Panosian Dunavan, and Jared Diamond, in *Improving Food Safety Through a One Health Approach: Institute of Medicine Workshop Summary* (National Academies Press, 2012). "The Origin of the Variola Virus," Igor V. Babkin and Irina N. Babkina, *Viruses* 7, no. 3 (March 10, 2015). *UNEP Frontiers 2016 Report: Emerging Issues of Environmental Concern*, UNEP Division of Early Warning and Assessment (United Nations Environment Programme, 2016). "Origins of the 1918 Pandemic: Revisiting the Swine 'Mixing Vessel' Hypothesis," Martha I. Nelson et al., *American Journal of Epidemiology* 187, no. 12 (2018). "The 'One Health' Concept Must Prevail to Allow Us to Prevent Pandemics," Eric Muraille and Jacques Godfroid, *Global Biodefense* (October 19, 2020). *Workshop Report on Biodiversity and Pandemics of the Intergovernmental Platform on Biodiversity and Ecosystem Services*, Peter Daszak et al. (IPBES, 2020).

"Anthrax Case Linked to Drumming Circle, New Hampshire Officials Say," Abby Goodnough, *The New York Times* (December 29, 2009). "Pneumonic Plague in a Dog and Widespread Potential Human Exposure in a Veterinary Hospital, United States," Paula A. Schaffer et al., *Emerging Infectious Diseases* 25, no. 4 (April 2019).

World at Risk: The Report of the Commission on the Prevention of WMD

Proliferation and Terrorism, Commission on the Prevention of Weapons of Mass Destruction Proliferation and Terrorism (Vintage Books, 2008).

"Agricultural Biological Weapons Threats: Food Safety, Security, and Emergency Preparedness," Jon Wefald, president, Kansas State University, testimony to the U.S. Senate's Emerging Threats Subcommittee (October 27, 1999). "Responding to the Threat of Agroterrorism: Specific Recommendations for the United States Department of Agriculture," Anne Kohnen, *BCSIA Discussion Paper 2000–29* (Belfer Center for Science and International Affairs, October 2000). "Found: The Islamic State's Terror Laptop of Doom," Harald Doornbos and Jenan Moussa, *Foreign Policy* (August 28, 2014). "A Threat to the Food System," Tom Daschle and Richard B. Myers, *U.S. News & World Report* (October 17, 2016). "Agroterrorism and Biosecurity Focus of Bipartisan Policy Center Event," *Global Biodefense* (November 2, 2016). "It's Alarmingly Easy for Terrorists to Contaminate Our Food Supply," Hank Parker, *New York Post* (January 7, 2017). "The Pentagon Is Studying an Insect Army to Defend Crops. Critics Fear a Bioweapon," Joel Achenbach, *The Washington Post* (October 4, 2018). "Reaping What You Sow: The Case for Better Agroterrorism Preparedness," Stevie Kiesel, *The Pandora Report* (February 20, 2020).

"Estimates of Inbreeding and Relationship Among Registered Holstein Females in the United States," C. W. Young and A. J. Seykora, *Journal of Dairy Science* 79, no. 3 (March 1996). "The Perfect Milk Machine: How Big Data Transformed the Dairy Industry," Alexis C. Madrigal, *The Atlantic* (May 1, 2012). "Farmers Business Network Is Disrupting How Seed Is Labeled and Sold," Gil Gullickson, *Successful Farming* (October 22, 2018). "What Farmers Can Expect from the Latest Round of Seed Company Mergers," Gil Gullickson, *Successful Farming* (June 27, 2019).

"Plum Island's Fate? Possibilities Abound," John Rather, *The New York Times* (September 4, 2005). "Secret Sanctions Revealed Against University Hosting $1.25 Billion Biolab," Alison Young, *USA Today* (August 4, 2015). "A Fortress of Bio Defense Research in Kansas," Amy Bickel, *The Hutchinson News* (November 12, 2017). "The Other Manhattan Project," Laura Ziegler and Brian Grimmett, *Harvest Public Media* (April 1, 2019).

"Withstanding High Wind Impacts on Biocontainment Facilities," Eugene Cole, presentation at the 59th Annual Biological Safety Conference (October 5, 2016). "Safeguarding American Agriculture in a Globalized World," Richard B. Myers, president, Kansas State University, Statement for the Record, Hearing of the United States Senate Agriculture Committee (December 13, 2017).

National Bio- and Agro-Defense Facility (NBAF) Feasibility Study, U.S. Department of Homeland Security (August 24, 2007). "DHS Lacks Evidence to Conclude That Foot-and-Mouth Disease Research Can Be Done Safely on the U.S. Mainland," Nancy Kingsbury, Testimony Before the Subcommittee on Oversight and Investigations, Committee on Energy and Commerce, House of Representatives (Government Accountability

Office, May 22, 2008). *Site-Specific Biosafety and Biosecurity Mitigation Risk Assessment: Final Report*, chair Ronald M. Atlas, National Academy of Sciences Committee, Department of Homeland Security (October 2010). *Evaluation of the Updated Site-Specific Risk Assessment for the National Bio- and Agro-Defense Facility in Manhattan, Kansas*, Committee on the Evaluation of the Updated Site-Specific Risk Assessment for the National Bio- and Agro-Defense Facility in Manhattan, Kansas (National Academies Press, 2012).

Texas Bio- & Agro-Defense Consortium v. United States of America, United States Court of Federal Claims, No. 09–255C (2009).

"Immigrant Women and Consumer Protest: The New York Kosher Meat Boycott of 1902," Paula E. Hyman, in *The American Jewish Experience*, 2nd ed., ed. Jonathan D. Sarna (Holmes & Meier, 1997).

Origin of the UK Foot and Mouth Disease Epidemic in 2001, Department for Environment, Food and Rural Affairs (June 2002). "Faulty Pipe Blamed for UK Foot and Mouth Outbreak," Andy Coghlan, *New Scientist* (September 7, 2007). "When Foot-and-Mouth Disease Stopped the UK in Its Tracks," Claire Bates, *BBC News Magazine* (February 17, 2016). "The Risks of Building Too Many Bio Labs," Elisabeth Eaves, *The New Yorker* (March 18, 2020).

"Iowa Searching for Help with Millions of Dead Chickens," Donnelle Eller, *The Des Moines Register* (May 17, 2015). "Midwest Farmers Rush to Dispose of Chickens Killed to Contain Avian Flu," Peggy Lowe, NPR (May 20, 2015). "The Looming Threat of Avian Flu," Maryn McKenna, *The New York Times Magazine* (April 13, 2016).

"Deadly Pig Virus May Have Sneaked into US on Reusable Bags," Maryn McKenna, *National Geographic* (October 8, 2015). "20,000 Pigs Culled Amid African Swine Fever Outbreak," *South China Morning Post* (August 23, 2018). "Build a Wall, Wild Boar Will Fall: Denmark Erects Barrier to Keep Out German Pigs," Emily Schultheis, *The Guardian* (January 28, 2019). "Half-Life of African Swine Fever Virus in Shipped Feed," Ana M. M. Stoian et al., *Emerging Infectious Diseases* 25, no. 12 (December 2019). "Chinese Criminal Gangs Spreading African Swine Fever to Force Farmers to Sell Pigs Cheaply So They Can Profit," Liu Zhen, *South China Morning Post* (December 14, 2019). "China Responds Slowly, and a Pig Disease Becomes a Lethal Epidemic," Keith Bradsher and Ailin Tang, *The New York Times* (December 17, 2019). "China Flight Systems Jammed by Pig Farm's African Swine Fever Defences," Mandy Zuo, *South China Morning Post* (December 20, 2019). "A 12-Storey Pig Farm: Has China Found the Way to Tackle Animal Disease?," Michael Standaert and Francesco De Augustinis, *The Guardian* (September 18, 2020).

"Pest Control in the Public Interest: Crop Protection in California," Brian P. Baker, *UCLA Journal of Environmental Law and Policy* 8, no. 31 (1998). "The Legacy of Charles Marlatt and Efforts to Limit Plant Pest Invasions," Andrew M. Liebhold and Robert L. Griffin, *American*

Entomologist 62, no. 4 (Winter 2016). "How the Gypsy Moth Came to America," Debbie Hadley, *ThoughtCo* (November 16, 2018). "This Plane Wasn't Snooping on Protesters in Los Angeles, It Was Dropping Irradiated Bugs," Joseph Trevithick, *The War Zone* (June 2, 2020).

"Quarantine Ordered to Stop Spread of Bee Mite," Don Kendall, Associated Press (April 6, 1988). "The Super Bowl of Beekeeping," Jaime Lowe, *The New York Times Magazine* (August 15, 2018). "Breeders Toughen Up Bees to Resist Deadly Mites," Erik Stokstad, *Science* (July 25, 2019). "Beekeepers Seek Resistance to the Honeybee's Most Fearsome Enemy," Paige Embry, *Undark* (October 1, 2019). "'Like Sending Bees to War': The Deadly Truth Behind Your Almond Milk Obsession," Annette McGivney, *The Guardian* (January 8, 2020).

"Jet-Setting Pets Get a New Place to Be Pampered at Kennedy Airport," Jane L. Levere, *The New York Times* (March 21, 2017). "A $65 Million 'Animal Terminal' at Kennedy Airport Sits Empty," Charles V. Bagli, *The New York Times* (January 24, 2018). "The ARK at JFK Expands Handling Capabilities into Medical Field," Rachelle Harry, *Air Cargo News* (June 5, 2020).

"Johnny Depp's Dogs Face Death in Australia," BBC News (May 14, 2015). "Strict Aussie Quarantine Laws Forced Equestrian Events for the 1956 Melbourne Olympics to Be Held in Stockholm," Troy Lennon, *The Daily Telegraph* (Sydney) (June 10, 2016). "Irreplaceable Plant Specimens an 'Obscene' Loss After Being Incinerated in Quarantine Flub," Christopher Mele and Aurelien Breeden, *The New York Times* (May 12, 2017).

"The French 'Bromance' Tree Could Be in Quarantine for Two Years," Betsy Klein, CNN (May 1, 2018). "Trump and Macron's Friendship Tree Dies After Being Put into Quarantine," Basit Mahmood, *Metro* (June 10, 2019).

"Quarantine Risks Imposed by Overseas Passengers," Edmund Sheridan, *New Zealand Journal of Forestry Science* 19, no. 2/3 (1989). "Red Menace: Stop the Ug99 Fungus Before Its Spores Bring Starvation," Brendan I. Koerner, *Wired* (February 22, 2010). "Virulent Wheat Fungus Invades South Africa," Natasha Gilbert, *Nature* (May 26, 2010). "Escalating Threat of Wheat Rusts," Mogens Støvring Hovmøller, Stephanie Walter, and Annemarie Fejer Justesen, *Science* 329, no. 5990 (July 23, 2010). "The Barberry Eradication Program in Minnesota for Stem Rust Control: A Case Study," Paul D. Peterson, *Annual Review of Phytopathology* 56 (August 2018).

"Spotswood Quarantine Centre to Be Absorbed into $379 Million Super Centre in Mickleham," Fiona O'Doherty, *Herald Sun* (Melbourne) (August 25, 2015). "How 139 Flocks of Sentinel Chickens Help Keep You Safe from Deadly Disease," Chris Haire, *Orange County Register* (March 29, 2018).

"Pest Risk Assessment of Insects in Sea Cargo Containers," Mark A. Stanaway et al., *Australian Journal of Entomology* 40, no. 2 (April 2001).

"Barrow Island Quarantine: Beyond Best Practice," Chevron Corporation (2015). "Zero-Tolerance Biosecurity Protects High-Conservation-Value Island Nature Reserve," Simon J. McKirdy et al., *Scientific Reports* 7, no. 772 (April 10, 2017).

7. A Million Years of Isolation

We first interviewed the U.S. Department of Energy geophysicist Abraham Van Luik by telephone, before meeting him in person at the Waste Isolation Pilot Plant in Carlsbad, New Mexico. Permission to take photographs inside WIPP was granted to us by a Department of Energy public affairs manager; in an email from Van Luik dated August 22, 2012, we learned that we were the last civilian visitors permitted to take unrestricted photos of the facility. Our final meeting and interview with Van Luik took place at a nuclear-waste disposal conference in Phoenix, Arizona, called Waste Management 2016, or WM2016. Sadly, Van Luik died a mere four months later, on July 9, 2016. Van Luik's blog, *Thoughts and Places*, was hosted online at thoughtsandplaces.org; it is now available only through internet archives.

"One Million Years of Isolation: An Interview with Abraham Van Luik," Geoff Manaugh and Nicola Twilley, *BLDGBLOG* (November 2, 2009).

This chapter benefited from interviews with other Department of Energy personnel, including Bobby St. John and Roger Nelson. We also spoke with Matt McCormick of Kurion about that firm's GeoMelt waste-vitrification technology; Kurion was acquired by Veolia shortly after our interview, in 2016.

A vast amount of material has been written about deep-geologic waste repositories. WIPP itself is the subject of voluminous materials archived online by the Department of Energy; these materials include more than one hundred thousand pages of licensing applications, environmental reviews, and technical reports. Of particular use to us were the following:

Waste Isolation Pilot Plant Land Withdrawal Act, Public Law 102–579, 106 Stat. 4777, United States Congress (1992).

40 CFR Part 191: *Environmental Radiation Protection Standards for the Management and Disposal of Spent Nuclear Fuel, High-Level and Transuranic Radioactive Waste*, Environmental Protection Agency (1993).

40 CFR Part 194: *Criteria for the Certification and Recertification of the Waste Isolation Pilot Plant's Compliance with the 40 CFR Part 191 Disposal Regulations*, Environmental Protection Agency (1996).

Waste Isolation Pilot Plant, *Title 40 CFR Part 191 Compliance Certification Application* (1996)—in particular, Appendix PIC ("Passive Institutional Controls Conceptual Design Report"), Appendix EPIC ("Effectiveness of Passive Institutional Controls in Reducing Inadvertent Human Intrusion into the Waste Isolation Pilot Plant for Use in Performance

Assessments"), and Appendix PEER 8 ("Passive Institutional Controls Peer Review").

"Communication Measures to Bridge Ten Millennia," Thomas A. Sebeok, Technical Report, Office for Nuclear Waste Isolation (April 1984). "Reducing the Likelihood of Future Human Activities That Could Affect Geologic High-Level Waste Repositories," Human Interference Task Force (May 1984). "Lebende Detektoren und komplementäre Zeichen: Katzen, Augen und Sirenen" ("Living Detectors and Complementary Signs: Cats, Eyes and Sirens"), Françoise Bastide and Paolo Fabbri, *Zeitschrift für Semiotik* (1984). "How Will Future Generations Be Warned?," U.S. Department of Energy, Carlsbad Field Office, Waste Isolation Pilot Plant (January 2003). "Permanent Markers Implementation Plan," John Hart and Associates (August 2004). "A Conceptual World Information Library (WIL) and Land Use Information System (LUIS)," Russell Patterson, Mindy Toothman, John Callicoat, and Tom Klein, paper delivered at WM2016, March 6–10, 2016, in Phoenix, Arizona.

"Worst Case Credible Nuclear Transportation Accidents: Analysis for Urban and Rural Nevada," Matthew Lamb and Marvin Resnikoff (Radioactive Waste Management Associates) and Richard Moore (2001). "Microbes Help Lock Up Nuclear Waste," Sam Wong, *New Scientist* (April 15, 2017). "Is Fukushima Doomed to Become a Dumping Ground for Toxic Waste?," Peter Wynn Kirby, *The Guardian* (March 16, 2018). "Italian Scientist Eyes Plant Growth at Nuclear Repository," Associated Press (December 15, 2018). "Biology @WIPP: Life Begins at 250,000,000 Years," Department of Energy press release, "Science at WIPP" (2019). It is worth noting the DOE's description of genetic research performed inside WIPP's North Experimental Area, complete with the exclamation point in the original text: "Scientists also have managed to cultivate bacteria from 250-million-year-old spores found in WIPP salt crystals—it's all similar to the plot of the movie *Jurassic Park*!" The resulting work, the DOE adds, has "pushed the envelope for resurrecting living things."

Review of the Department of Energy's Plans for Disposal of Surplus Plutonium in the Waste Isolation Pilot Plant, National Academies of Sciences, Engineering, and Medicine (National Academies Press, 2020).

Principles and Standards for the Disposal of Long-Lived Radioactive Wastes, Neil Chapman and Charles McCombie, Waste Management Series, vol. 3 (2003). *Signs of Danger: Waste, Trauma, and Nuclear Threat*, Peter C. van Wyck (University of Minnesota Press, 2004). *Uncertainty Underground: Yucca Mountain and the Nation's High-Level Nuclear Waste*, ed. Allison M. Macfarlane and Rodney C. Ewing (MIT Press, 2006). *Too Hot to Touch: The Problem of High-Level Nuclear Waste*, William M. Alley and Rosemarie Alley (Cambridge University Press, 2013).

"The Beishan Underground Research Laboratory for Geological Disposal of High-Level Radioactive Waste in China," Ju Wang, Liang Chen,

Rui Su, and Xingguang Zhao, *Journal of Rock Mechanics and Geotechnical Engineering* 10, no. 3 (June 2018). "China Earmarks Site to Store Nuclear Waste Deep Underground," Echo Xie, *South China Morning Post* (September 5, 2019).

"UN Agency Warns About Nuclear Scrap Contaminating Consumer Goods," *The New York Times* (October 11, 2008). "Nuclear Risks at Bed Bath & Beyond Show Dangers of Scrap," Jonathan Tirone and Andrew MacAskill, *Bloomberg* (March 19, 2012). "Chernobyl Bombshell: Radioactive Metal from Exclusion Zone Could Be 'Used in Phones Today,'" Callum Hoare, *Express* (June 13, 2019).

Radiological Release Event at the Waste Isolation Pilot Plant, February 14, 2014, U.S. Department of Energy Office of Environmental Management Accident Investigation Report (April 2015). "How the Wrong Cat Litter Took Down a Nuclear Waste Repository," Jessica Morrison, *Chemical & Engineering News* (May 15, 2017).

Michael Madsen's feature documentary *Into Eternity* (2010) is useful for its depiction of permanent geological disposal at Onkalo, as is Robert Macfarlane's book *Underland: A Deep Time Journey* (Hamish Hamilton, 2019).

8. All the Planets, All the Time

This chapter draws on a series of meetings, interviews, and tours with Judith Allton, Nick Bernadini, Penelope Boston, Catherine Conley, Gerhard Kminek, Paolo Nespoli, Lisa Pratt, and Kasthuri Venkateswaran. These interviews took place at NASA's Washington, D.C., headquarters; the NASA Johnson Space Center in Houston, Texas; the Jet Propulsion Laboratory in Pasadena; the European Space Agency in Noordwijk, the Netherlands; and Penelope Boston's home in New Mexico.

An essential source for this chapter was *When Biospheres Collide: A History of NASA's Planetary Protection Programs*, Michael Meltzer (NASA, 2011).

The Planetary Quarantine Program: Origins and Achievements, 1956–1973, Charles R. Phillips, NASA Special Publication 4902 (1974). "From Planetary Quarantine to Planetary Protection: A NASA and International Story," John D. Rummel, *Astrobiology* 19, no. 4 (2019).

1491: New Revelations of the Americas Before Columbus, Charles C. Mann (Alfred A. Knopf, 2005).

"Conquest and Population: Maya Demography in Historical Perspective," W. George Lovell and Christopher H. Lutz, *Latin American Research Review* 29, no. 2 (1994). "Spanish and Nahuatl Views on Smallpox and Demographic Catastrophe in the Conquest of Mexico," Robert McCaa, *Journal of Interdisciplinary History* 25, no. 3 (Winter 1995). "Skeletons in the Closet: The Smithsonian's Native American Remains and the NMAI," Karis Lee, *Boundary Stones* (May 21, 2020).

Perelandra, C. S. Lewis (Simon and Schuster, 1996; originally published 1943).

Emerging Infections: Microbial Threats to Health in the United States, ed. Joshua Lederberg, Robert E. Shope, and Stanley C. Oaks Jr., for the Institute of Medicine Committee on Emerging Microbial Threats to Health (National Academies Press, 1992). "Germs in Space: Joshua Lederberg, Exobiology, and the Public Imagination, 1958–1964," Audra J. Wolfe, *Isis* 93 (2002). "UFOs and Aliens Among Us," Library of Congress Digital Collections, *Finding Our Place in the Cosmos: From Galileo to Sagan and Beyond.*

"Apollo 12 Trailer Resurfaces at a Fish Farm?," John Peck, *The Huntsville Times* (August 11, 2007). "History Spotlight: Airstream's Place in Space," Airstream, *Live Riveted* (July 15, 2009). "50 Years Ago: Mobile Quarantine Facility Arrives in Houston," NASA (March 6, 2018).

Conference on Potential Hazards of Back Contamination from the Planets, National Research Council (National Academies Press, 1964). "Mission Operation Report: Apollo 11 Mission," NASA Report No. M-932-69-11 (June 24, 1969). *Biomedical Results of APOLLO*, Richard S. Johnston, Lawrence F. Dietlein, and Charles A. Berry (NASA Special Publication 368, 1975). *Where No Man Has Gone Before: A History of Apollo Lunar Exploration Missions*, W. David Compton (NASA Special Publication 4214, 1989). "25 Years of Curating Moon Rocks," Judy Allton, *Lunar News* 57 (July 1994). "Lessons Learned During Apollo Lunar Sample Quarantine and Sample Curation," J. H. Allton, J. R. Bagby Jr., and P. D. Stabekis, *Advances in Space Research* 22, no. 3 (1998). "Moon Rocks and Moon Germs: A History of NASA's Lunar Receiving Laboratory," Kent Carter, *Prologue* 33, no. 4 (Winter 2001). "The Case of the Stolen Moon Rocks: Last of Three NASA Interns Sentenced for Grievous Theft," Federal Bureau of Investigation (November 18, 2003). "Lunar Receiving Laboratory Project History," Susan Mangus and William Larsen, NASA/CR–208938 (June 2004). *Lunar Sample Allocation Guidebook*, Astromaterials Acquisition and Curation Office (NASA, June 2007). *The Quarantine and Certification of Martian Samples*, Committee on Planetary and Lunar Exploration, Space Studies Board (National Academies Press, 2002). "Extraterrestrial Exposure Quarantine Laws and Apollo 11," Derek T. Muller, *Excess of Democracy* (July 15, 2014). "This Month in NASA History: From the Moon to an Airstream Trailer," Kevin Wilcox, NASA Appel Knowledge Services (August 15, 2019).

Houston, Space City USA, Ray Viator (Texas A&M University Press, 2019).

Oral history transcripts from NASA Johnson Space Center Oral History Project: Charles A. Berry, interviewed by Carol Butler (Houston, Texas, April 29, 1999); Brock R. "Randy" Stone, interviewed by Sandra Johnson (Houston, Texas, October 18, 2006); and John K. Hirasaki, interviewed by Sandra Johnson (Houston, Texas, March 6, 2009).

Chasing the Moon, directed by Robert Stone, PBS (July 2019).

"Moon 'Genesis Rock' 4 Billion Years Old," John Noble Wilford (September 18, 1971). "Water in Lunar Anorthosites and Evidence for a Wet Early Moon," Hejiu Hui et al., *Nature Geoscience* 6 (2013).

"Life Below and Life 'Out There,'" Penelope Boston, *Geotimes* (August 2000). *Preventing the Forward Contamination of Mars*, National Research Council (National Academies Press, 2006). "Life on the Subsurface: An Interview with Penelope Boston," Geoff Manaugh and Nicola Twilley, *Venue* (July 23, 2014). "Meet the Martians," Nicola Twilley, *The New Yorker* (October 8, 2015). "Food on Enceladus, Old Faithful on Europa Strengthen Case for Finding Alien Life," Lee Billings, *Scientific American* (April 13, 2017). "ExoMars Finds New Gas Signatures in the Martian Atmosphere," European Space Agency (July 27, 2020).

"Spore UV and Acceleration Resistance of Endolithic *Bacillus pumilus* and *Bacillus subtilis* Isolates Obtained from Sonoran Desert Basalt: Implications for Lithopanspermia," James N. Benardini, John Sawyer, Kasthuri Venkateswaran, and Wayne L. Nicholson, *Astrobiology* 3, no. 4 (July 5, 2004). "Newfound Bacteria Fueled by Radiation," David Brown, *The Washington Post* (October 20, 2006). "Description of *Rummeliibacillus stabekisii* gen. nov., sp. nov.," Kasthuri Venkateswaran et al., *International Journal of Systematic and Evolutionary Microbiology* 59, pt. 5 (May 2009). "Whole-Genome Sequence of *Rummeliibacillus stabekisii* Strain PP9 Isolated from Antarctic Soil," Fábio Faria da Mota et al., *Genome Announcements* 4, no. 3 (May–June 2016).

"In Search of a Second Genesis," Paul Davies, *New Scientist* (February 8, 2006). "Keeping Alien Invaders at Bay," Paul Marks, *New Scientist* (April 28, 2007). "Did a Distant Solar System Send Life to Earth? A New Theory of How Space Debris Travels Increases the Likelihood That Star Systems Swap Biology," Jeffrey Kluger, *Time* (September 26, 2012). "Searching for Life on Mars Before It Is Too Late," Albert G. Fairén et al., *Astrobiology* 17, no. 10 (October 2017). International Committee Against Mars Sample Return (www.icamsr.org). "Swinging on a Star," Elizabeth Kolbert, *The New Yorker* (January 25, 2021).

"How Will Police Solve Murders on Mars?," Geoff Manaugh, *The Atlantic* (September 14, 2018). "A Crashed Israeli Lunar Lander Spilled Tardigrades on the Moon," Daniel Oberhaus, *Wired* (August 5, 2019). "Elon Musk's SpaceX Will 'Make Its Own Laws on Mars,'" Anthony Cuthbertson, *The Independent* (October 27, 2020).

"Treaty on Principles Governing the Activities of States in the Exploration and Use of Outer Space, Including the Moon and Other Celestial Bodies," U.S. Department of State, signed in Washington, London, and Moscow on January 27, 1967; entered into force on October 10, 1967.

"'My Sister Says I Am an Alien': A 9-Year-Old Applies to Be NASA's Planetary Protection Officer," Amy B. Wang, *The Washington Post* (August 6, 2017). *Review and Assessment of Planetary Protection Policy Development*

Processes, Committee on the Review of Planetary Protection Policy Development Processes, Space Studies Board, National Academy of Sciences (2018). *Report from NASA's Planetary Protection Independent Review Board*, NASA (2019). "Lacuna in the Updated Planetary Protection Policy and International Law," Patricia M. Sterns and Leslie I. Tennen, in *Life Sciences in Space Research* 23 (2019). *Assessment of the Report of NASA's Planetary Protection Independent Review Board*, Committee to Review the Report of the NASA Planetary Protection Independent Review Board, Space Studies Board (2020).

Endurance: My Year in Space, a Lifetime of Discovery, Scott Kelly (Alfred A. Knopf, 2017).

"Headed to Space? Tips for Travel," Mark Bloom, *MedPage Today* (December 19, 2012). "Apollo 7: NASA's First Mini-mutiny in Space," Bill Andrews, *Discover* (October 9, 2018).

9. Algorithms of Quarantine

For this chapter, we attended a number of pandemic-simulation exercises: Airport Roles in Reducing Transmission of Communicable Diseases, March 2018, Washington, D.C., sponsored by the Transportation Research Board and the National Academies of Sciences; Clade X, May 2018, Washington, D.C., sponsored by the Johns Hopkins Center for Health Security; National Summit on Pandemic Preparedness, May 2019, Omaha, Nebraska, sponsored by the National Center for State Courts, the State Justice Institute, and the University of Nebraska Medical Center; and Event 201, October 2019, New York City, sponsored by the Johns Hopkins Center for Health Security in partnership with the World Economic Forum and the Bill and Melinda Gates Foundation.

This chapter relies on interviews with Dirk Brockmann, Dr. Martin Cetron, Cris Collinsworth, Dr. Julie L. Gerberding, Dr. Tom Inglesby, Adam Kucharski, and Jill Stelfox.

The Great Indoors: The Surprising Science of How Buildings Shape Our Behavior, Health, and Happiness, Emily Anthes (Scientific American / Farrar, Straus and Giroux, 2020). *Super-Cannes*, J. G. Ballard (Picador, 2001). *Ubik*, Philip K. Dick (Doubleday, 1969). *Discipline and Punish: The Birth of the Prison*, Michel Foucault (Vintage Books, 1995). *Chaos: Making a New Science*, James Gleick (Penguin, 1987). *The Rules of Contagion: Why Things Spread—and Why They Stop*, Adam Kucharski (Basic Books, 2020).

"International Sanitary Convention for Aerial Navigation," *The American Journal of International Law* (January 1937). "Preventive Medicine in Relation to Aviation: President's Address," Harold Whittingham, *Proceedings of the Royal Society of Medicine* (December 12, 1938).

"Behaviors, Movements, and Transmission of Droplet-Mediated Respiratory Diseases During Transcontinental Airline Flights," Vicki Stover Hertzberg et al., *PNAS* (April 3, 2018). "An AI Epidemiologist Sent the

First Warnings of the Wuhan Virus," Eric Niiler, *Wired* (January 25, 2020). "Modelling Airport Catchment Areas to Anticipate the Spread of Infectious Diseases Across Land and Air Travel," Carmen Huber et al. (including Dr. Martin Cetron and Kamran Khan), *Spatial and Spatio-temporal Epidemiology* 36 (2021).

"The Hidden Geometry of Complex, Network-Driven Contagion Phenomena," Dirk Brockmann and Dirk Helbing, *Science* (December 13, 2013). "Pandemic Geography," Dirk Brockmann, *The Geographer* (Winter 2013–2014). "Global Connectivity and the Spread of Infectious Diseases," Dirk Brockmann, *Nova Acta Leopoldina*, no. 419 (2017).

"A Successful Launch to the International Space Station," Kenneth Chang, *The New York Times* (March 27, 2015).

"Palantir, a Data Firm Loved by Spooks, Teams Up with Britain's Health Service," *The Economist* (March 26, 2020). "FEMA Tells States to Hand Public Health Data Over to Palantir," Spencer Ackerman, *The Daily Beast* (May 21, 2020). "Does Palantir See Too Much?," Michael Steinberger, *The New York Times Magazine* (October 21, 2020). "The Path Forward: Combating COVID-19 with Palantir CEO Alexander Karp," *The Washington Post Live* (January 7, 2021).

"What We Can Learn from the Epic Failure of Google Flu Trends," David Lazer and Ryan Kennedy, *Wired* (October 1, 2015). "'Smart Thermometers' Track Flu Season in Real Time," Donald G. McNeil Jr., *The New York Times* (January 16, 2018). "This Thermometer Tells Your Temperature, Then Tells Firms Where to Advertise," Sapna Maheshwari, *The New York Times* (October 23, 2018). "Can This Thermometer Help America Reopen Safely?," Charlie Warzel, *The New York Times* (June 29, 2020). "What Negative Candle Reviews Might Say About the Coronavirus," Christopher Ingraham, *The Washington Post* (December 1, 2020).

"Voice-Based Determination of Physical and Emotional Characteristics of Users," Amazon Technologies Inc., United States Patent 10,096,319 B1 (October 9, 2018). "Amazon Introduces 'Distance Assistant,'" Brad Porter, *About Amazon* blog (June 16, 2020). "COVID-19 Artificial Intelligence Diagnosis Using Only Cough Recordings," Jordi Laguarta, Ferran Hueto, and Brian Subirana, *IEEE Journal of Engineering in Medicine and Biology* (September 29, 2020).

"How Will Real-Time Tracking Change the N.F.L.?," Nicola Twilley, *The New Yorker* (September 10, 2015). "London Marathon Becomes Elite-Only Race Running Laps in St. James's Park," Sean Ingle, *The Guardian* (August 6, 2020).

"Chinese Citizens Turn to Virus Tracker Apps to Avoid Infected Neighborhoods," Reuters (February 3, 2020). "As Coronavirus Surveillance Escalates, Personal Privacy Plummets," Natasha Singer and Choe Sang-Hun, *The New York Times* (March 23, 2020). "Countries Are Using Apps and Data Networks to Keep Tabs on the Pandemic," *The Economist* (March 26, 2020). "Apple, Google Debut Major Effort to Help People Track If They've

Come in Contact with Coronavirus," Tony Romm, Drew Harwell, Elizabeth Dwoskin, and Craig Timberg, *The Washington Post* (April 10, 2020). "The Hot New Covid Tech Is Wearable and Constantly Tracks You," Natasha Singer, *The New York Times* (November 15, 2020).

Epilogue: Until Proven Safe

The epilogue draws on interviews with Dr. Sylvie Briand, Dr. Martin Cetron, Col. Dr. Matthew Hepburn, and Erin Westgate.

Acknowledgments

Our journey through quarantine began in the autumn of 2009, long before the specific idea for *Until Proven Safe* took shape. For four months in New York City that autumn, we hosted weekly conversations with a small group of architects, artists, designers, and writers to explore what the concept of quarantine—isolation and risk, exposure and uncertainty—might mean in their respective fields. The results of that research, from short fiction to theatrical set designs, were then displayed at Storefront for Art and Architecture in spring 2010 under the title *Landscapes of Quarantine*. We thank Joe Alterio, Front Studio (Yen Ha and Michi Yanagishita), Scott Geiger, Katie Holten, Jeffrey Inaba, Ed Keller, Mimi Lien, Richard Mosse, Daniel Perlin, Thomas Pollman, Kevin Slavin, Brian Slocum, Smudge Studio (Elizabeth Ellsworth and Jamie Kruse), and Amanda and Jordan Spielman for participating in those early, formative conversations. We also thank Glen Cummings for his superb exhibition design and Joseph Grima, the then director of Storefront, for allowing us to curate that exhibition in the first place.

Throughout years of travel, beginning in 2016, we benefited at every turn—at every port, harbor, and library—from the intellectual camaraderie and personal generosity of dozens. We owe thanks to the following people, in particular: Judith Allton, Joanne Andreadis, David Barnes, Dr. Mark Barnes, Alison Bashford, Dr. Georges Benjamin, James Benardini, Dr. Luigi Bertinato, Penelope Boston, Dr. Sylvie Briand, Dirk Brockmann, Dr. Clive Brown, Birsen Bulmuş, Dr. John Cachia, Dr. Martin Cetron, Guillaume Chabot-Couture, Dr. Ted Cieslak, Eugene Cole, James Colgrove, Cris Collinsworth, Catherine Conley, Stephanie Dahl, Ugo Del Corso, Richard DeLighter, Tatyana Eatwell, Jennifer Elsea, Mike Famulare, Nicolina Farrugia, Gerolamo Fazzini, Debbie Felton, Matthew Fulks, Dr. Julie L. Gerberding, Wayt Gibbs, Lawrence Gostin, Owen Guo, Paul Hadley, Dr. Margaret Hamburg, John Henneman, Col. Dr. Matthew Hepburn, Kaci Hickox, Stephen Higgs, Nancy Hollander,

Dr. Tom Inglesby, Dr. Sir Mike Jacobs, Michele Jacobsen, Yue Jin, Dr. Papy Katabuka, Shahryar Kianian, Allison Klajbor, Gerhard Kminek, James Kolmer, Dr. Phyllis Kozarsky, Adam Kucharski, Heather Lake, Elizabeth Landau, Dr. Patrick LaRochelle, Dr. Herbert Lenicker, Rachel Lookadoo, Krista Maglen, Simon McKirdy, Jill Morgan, Angela Munari, (Ret.) General Richard B. Myers, Nathan Myhrvold, Roger Nelson, Paolo Nespoli, Snježana Perojević, Noa Pinter-Wollman, Lisa Pratt, Fausto Pugnaloni, Sara Redstone, Jonathan Y. Richmond, Matthew Rouse, Edward Said, Bobby St. John, (Ret.) Chief Engineer Todd T. Semonite, James Stack, Jay Stanley, Jill Stelfox, Les Szabo, Ron Trewyn, Bahar Tuncgenc, Dr. Patrick Ucama, Abraham Van Luik, Denis Vandervelde, Marty Vanier, Dr. Anthony Vassallo, Mitch Vega, Kasthuri Venkateswaran, Erin Westgate, and Paige Williams.

While conducting much of the research and reading for this book, we relied on the resources of the Wellcome Library in London, the Querini Stampalia Library in Venice, and Sci-Hub online.

We owe a huge thanks to our editors. In the United States, Sean McDonald displayed both patience and prescience in sending us off to research six hundred years of quarantine. Our entire team at MCD / Farrar, Straus and Giroux worked tirelessly to see this book through to publication. In the U.K., Georgina Morley offered us an early expression of interest and encouragement that helped the book's initial form take shape. Our agent, Nathaniel Jacks, was a source of support throughout. Those elements previously published by *The New Yorker* and *The New York Times* benefited immensely from the editorial attention of Anthony Lydgate and Alan Burdick, respectively. Wayne Chambliss offered ongoing intellectual support and an extremely helpful close read.

Any remaining errors in the text, of course, are our own.

Geoff Manaugh would also like to thank Mónica Belevan for inviting him to join the COVID-19 prepper list described in chapter 1; Benjamin Bratton, Nicolay Boyadjiev, and Olga Tenisheva at the Strelka Institute for multiple invitations to lecture on quarantine for their students in Moscow, Russia; the Canadian Centre for Architecture for an invitation to deliver a talk called "Cities of the CDC," on architecture, medical infrastructure, and the future of quarantine, back in 2009; and, of course, Nicola Twilley for everything so far and many more decades together to come.

Nicola Twilley would also like to thank her *Gastropod* cohost Cynthia Graber for her endless patience; Victoria Wade for her commitment to worrying about deadlines; Ellie Robins and Lizzie Prestel for the accountability; Siri Carpenter and Christie Aschwanden for their advice and encouragement; Anne Roughley for a delightful picnic at Q Station; and, of course, Geoff Manaugh for all the good ideas, from getting married to writing this book together.

Index

A NOTE ABOUT THE AUTHORS

Geoff Manaugh is the author of the *New York Times* best-seller *A Burglar's Guide to the City*, as well as the architecture and technology website *BLDGBLOG*. He regularly writes for *The New York Times Magazine*, *The Atlantic*, *The New Yorker*, *Wired*, and many other publications. Nicola Twilley is the cohost of the award-winning podcast *Gastropod*, which looks at food through the lens of science and history, and is a frequent contributor to *The New Yorker*. Manaugh and Twilley live in Los Angeles.